Mainstreaming Natural Capital and Ecosystem Services into Development Policy

This book highlights the latest advances in the science and practice of using ecosystem services to inform decisions for economic development in the context of the developing countries.

The development of the ecosystem services paradigm has enhanced our understanding of natural capital as an indispensable form of capital asset along with produced and human capital. This book addresses what could be the possible pathways to mainstream natural capital assets into development policies and what is currently known about the economic values of ecosystem services. A series of innovative tools to help policy makers and planners to account for natural capital and ecosystem services in sectoral and macroeconomic policies has been explored and their application at the national and regional scale has been demonstrated. Several detailed case studies are presented in which the understanding of ecosystem services values has successfully informed decisions, including examples from Chile, South Africa, Tanzania, Trinidad and Tobago, Vietnam and the Aral Sea in Central Asia. These provide the critically important insights, lessons learned and means and mechanisms for policy makers to incentivize protection and discourage degradation of ecosystems and the services they provide.

Mainstreaming Natural Capital and Ecosystem Services into Development Policy is designed to help decision makers at all levels, including governments, businesses, multilevel development banks and individuals to integrate ecosystems and their services into their decision making.

Pushpam Kumar is Chief Environmental Economist and Senior Economic Advisor, UN Environment Programme. He works on the issues of Inclusive Wealth Index, Natural Capital, Climate Change, Environmental Burden of Disease, Human Capital and Sustainable Development Goals (SDGs).

T0175015

Mainstreaming Natural Capital and Ecosystem Services into Development Policy

Edited by Pushpam Kumar

LONDON AND NEW YORK

First published 2019
by Routledge
2 Park Square, Milton Park, Abingdon, Oxon OX14 4RN

and by Routledge
605 Third Avenue, New York, NY 10017

First issued in paperback 2021

Routledge is an imprint of the Taylor & Francis Group, an informa business

ISBN 13: 978-0-367-77698-5 (pbk)
ISBN 13: 978-1-138-69311-1 (hbk)

British Library Cataloguing-in-Publication Data
A catalogue record for this book is available from the British Library

Library of Congress Cataloging-in-Publication Data
A catalog record for this book has been requested

Typeset in Times New Roman
by Apex CoVantage, LLC

Contents

Figures

Tables

Contributors

Babatunde Abidoye, Climate Change Adaptation, United Nations Development Programme; The Centre for Environmental Economics and Policy in Africa; Environmental Science and Policy, George Mason University, Fairfax, Virginia, USA; The School of Public Management and Administration, University of Pretoria, Pretoria, South Africa.

John Agard, Department of Life Sciences, University of the West Indies, St. Augustine Campus, Port of Spain, Trinidad and Tobago.

Carlos Muñoz Brenes, Department of Natural Resources and Society, University of Idaho, Moscow, USA.

Bernardo Broitman, Department of Sciences, Faculty of Liberal Arts, Adolfo Ibañez University, Viña del Mar; Center for Advanced Studies in Arid Zones, La Serena, Chile.

Lorena Bugueño, Municipality of Salamanca; The Centre of Advanced Studies on Arid Zones, La Serena, Chile.

Vera Camacho-Valdez, El Colegio de la Frontera Sur (ECOSUR), Periférico Sur s/n, Colonia María Auxiliadora, San Cristóbal de las Casas, Mexico.

Lena Dempewolf, Department of Life Sciences, University of the West Indies, St. Augustine Campus, Port of Spain, Trinidad and Tobago.

Keisha Garcia, Department of Life Sciences, University of the West Indies, St. Augustine Campus, Port of Spain, Trinidad and Tobago.

Cristian Geldes, Department of Economics and Business, Universidad Alberto Hurtado, La Serena, Chile.

Davide Geneletti, Department of Civil, Environmental and Mechanical Engineering, University of Trento, Trento, Italy.

Andrea Ghermandi, Department of Natural Resources and Environmental Management, University of Haifa, Haifa, Israel.

Leticia González-Silvestre, Municipality of San Pedro de Atacama; The Centre of Advanced Studies on Arid Zones, La Serena, Chile.

Tran Thi Thu Ha, Research Institute for Forest Ecology and Environment, Vietnam Academy of Forest Science, Ha Noi, Vietnam.

Ngo Chi Hung, Department of Environmental Protection, Ca Mau Department of Natural Resources and Environment, Ha Noi, Vietnam.

Jana Juhrbandt, Consultant for Environmental Economics and Natural Resources Protection, Göttingen Lower Saxony, Germany.

Tran Trung Kien, Disaster Management Center (DMC), Ministry of Agriculture and Rural Development (MARD), Ha Noi, Vietnam.

Pushpam Kumar is Chief Environmental Economist and Senior Economic Advisor, UN Environment Programme, Nairobi, Kenya.

Anil Markandya, Basque Center for Climate Change, Bilbao, Spain.

Eric Mungatana, The Centre for Environmental Economics and Policy in Africa, Pretoria, South Africa, Switzerland.

Jeanne Nel, Wageningen University & Research, Wageningen; Sustainability Research Unit, Nelson Mandela University, Port Elizabeth, South Africa.

Kim Thi Thuy Ngoc, Institute of Strategy and Policy on Natural Resources and Environment, Ministry of Natural Resources and Environment, Vietnam.

Carl Obst, Institute for the Development of Environmental-Economic Accounting (IDEEA), Melbourne, Australia.

Michael Parsons, Ministry of Natural Resources and Environment, Ha Noi, Vietnam.

Le Thi Le Quyen, Institute of Strategy and Policy on Natural Resources and Environment, Ministry of Natural Resources and Environment, Ha Noi, Vietnam.

Maurice Andres Rawlins, Department of Life Sciences, University of the West Indies, St. Augustine Campus, Port of Spain, Trinidad and Tobago; The World Bank, Washington D.C, USA.

Belinda Reyers, Centre for Complex Systems in Transition (CST), Stellenbosch University, South Africa; Future Africa, University of Pretoria, Pretoria, South Africa.

Terry Roe, Department of Applied Economics, University of Minnesota, Saint Paul, USA.

Sonia Salas, Department of Psychology, Universidad de La Serena, La Serena, Chile.

Nadia Sitas, Centre for Complex Systems in Transition (CST), Stellenbosch University, Stellenbosch; Natural Resources and the Environment Unit, Council for Scientific and Industrial Research (CSIR), Pretoria, South Africa.

Rodney Smith, Department of Applied Economics, University of Minnesota, Saint Paul, USA.

Eric Sproles, Department of Earth Science, Montana State University, Bozeman, USA; and The Centre of Advanced Studies on Arid Zones, La Serena, Chile.

Pavan Sukhdev, Green Initiatives for a Smart Tomorrow (GIST) Advisory; WWF International, Gland, Switzerland.

Nguyen Van Tai, Viet Nam Environment Administration (VEA), Ministry of Natural Resources and Environment, Vietnam.

Shirley Murillo Ulate, Individual Consultant on Environmental Services and Management of Natural Resources, Monteverde, Costa Rica.

Kaavya Varma, UN Environment and Sustainable India Finance Facility (SIFF), New Delhi, India.

Gregg Michael Verutes, Faculty of Political and Social Sciences, Department of Applied Economics, University of Santiago de Compostela, Santiago de Compostela, Spain.

Craig Weideman, Ya'axche Conservation Trust, Punta Gorda, Belize; The Centre of Advanced Studies on Arid Zones, La Serena, Chile.

Antonia Zambra, Department of Sociology, Universidad Diego Portales, Santiago, Chile; The Centre of Advanced Studies on Arid Zones, La Serena, Chile.

Acknowledgements

The editor would like to express sincere thanks to Amelia Holmes for her valuable contribution and assistance in overseeing this publication. We would also like to gratefully acknowledge Ersin Essen for his coordination of the GEF-supported PROECOSERV and Maxwell Gomera as director of Biodiversity and Ecosystems Branch, UN Environment.

This publication would not have been possible without the generous support of the Global Environmental Facility (GEF), The United Nations Programme on Reducing Emissions from Deforestation and Forest Degradation (UN-REDD), and UN Development Accounts, New York.

1 Introduction

Pushpam Kumar

Nothing is more powerful than an idea whose time has come.

Victor Hugo

When the 19th century French novelist Victor Hugo wrote this, he was simply trying to convey the potency and universality of a relevant idea in the era of industrial revolution, which was characterized by mechanized production.

The measurement and management of "natural capital" is one such idea whose time has come in the fourth industrial revolution, which is characterized by the digital world, genetics, robotics, the Internet and 5G networks. If anything, the concept of natural capital is more relevant than ever in 2019, when we are devising the means to achieve the Sustainable Development Goals.

"Natural capital" refers to assets that occur in nature and can be used for economic production or consumption. Naturally occurring assets provide use-benefits through the provision of energy and raw materials that are used, or that one day could be used, in economic activity. They are subject to depletion through human use. Natural assets fall into four categories: mineral and energy resources, soil resources, water resources and biological resources.

The need for a new economics for sustainable development exclusively focusses on the prosperity of people and planet. This calls for mainstreaming various elements of natural capital into economic and decisions of the society and economy.

There are various pathways through which mainstreaming can be understood better. The quintessential features of this mainstreaming run through following (Polasky et al., 2019):

1   Valuation of Natural Capital
2   Recognizing uncertainty
3   Capturing uncertainty and long-term consequences
4   Recognising the inertia and drivers of behavioural changes
5   Relevance of methodological pluralism
6   Role of innovation and equity in the approach and tools used

To date, society has not developed an indicator comprehensive enough to capture the change in natural capital over time. Many elements of natural capital are downward trend in their natural capital over the past two decades. Forest resources are declining in 85 countries. Renewable resources in general are declining in more than 100 countries (Managi and Kumar, 2018).

A measure of national wealth that excludes natural capital depreciation likely exaggerates the actual increase in an economy's wealth over time, especially in those countries where accumulation of other forms of wealth is failing to compensate for diminishing natural capital. This suggests that income and wealth inequality may be worsening in rich countries and in the global economy in general.

It is not surprising that at the UN Environment Assembly in May 2016, the nations of the world, in an unprecedented move, unanimously passed a resolution calling for the sustainable management of natural capital for sustainable development and poverty eradication. This was due to the culmination of the urgency expressed by global communities over the last decade.

Today we are in the fourth stage of the industrial revolution, which has driven productivity and efficiency to a new level. Unfortunately, this revolution does not reduce the burden on natural capital, especially in the poorest parts of the world – sub-Saharan Africa, Amazonia and South Asia – where 896 million people still face extreme poverty and rely on common water bodies, grazing land and biodiversity to survive. On an everyday basis, cities are flooding and schools are being closed due to air pollution. If we continue with business as usual, plastics will soon exceed the fish stock in our oceans.

Global and national modelling studies suggest that yields of major cereals will decline if temperatures continue to rise, especially in tropical countries. Water scarcity threatens agricultural growth in large parts of the world. Declines in natural capital – especially the loss of biodiversity, including critical crop pollinators, and a reduction in soil quality – will have substantial impacts on global food security.

We need better measurement and credible valuation of natural capital in order to integrate these natural assets into our everyday production and consumption decisions.

Global consensus is emerging on the need to measure and value natural capital. This can be seen in the following declarations and decisions worldwide.

Gaborone Declaration 2012, in which ten African countries undertook the pledge to "ensure that the contributions of natural capital to sustainable economic growth, maintenance and improvement of social capital and human well-being are quantified and integrated into development and business practice".

African ministers of the environment (in AMCEN under Decision 15/1 – Sustainably harnessing Africa's natural capital in the context of Agenda 2013) agreed to take measures at the national, regional and international levels to combat the illicit trade of Africa's natural resources, reverse illicit flows, and ensure the restitution of such resources to their countries of origin.

They agreed to put in place appropriate policies and institutional measures, taking into consideration the findings of the report entitled *Africa's Adaptation Gap 2: Bridging the Gap – Mobilizing Sources*, so as to: harness the full potential of Africa's rich natural resource endowments; ensure that the use of natural resources benefits the countries that possess them in an inclusive manner; create value addition in sustainably managing natural resources while ensuring the protection of ecosystems and minimizing environmental degradation; and promote sustainable consumption and production patterns; finally, they also agreed to strengthen international cooperation to promote research and development and technological innovations, as well as capacity development for sustainably harnessing the continent's natural resources.

The Fourth Environment Ministers' Meeting in Asia in 2015 (attended by China, Lao PDR, Myanmar, Thailand and Vietnam) stressed the need for assessment, measurement, accounting and developing linkages of natural capital with economic policies of governments, business and industry.

The Sustainable Development Goals will appear as the common thread but some core issues in measurement and accounting of natural capital will also get significant attention. In addition, other issues such as natural capital and equity and the possibility of substitution of natural capital by other types of capital will be discussed.

There will be engaging discussions on how to measure wealth and use it as a complement to conventional progress measurement.

Sustainability is mainly meeting the needs of the present without compromising the ability of future generations to meet their own needs (Brundtland et al., 1987), and degradation of natural capital affects sustainability adversely and places a growing burden on human well-being and development. Natural capital is an important share of total wealth which is determinant of development. Natural capital is of direct/indirect use or of use in both. The worth of a natural resource could be based on what is extracted from it or on its presence as a stock offering service.

Effectiveness of development can be affected significantly by degrading natural capital. Development is either directly or indirectly linked to natural capital, yet limited attention has been given by decision makers in discerning the role of healthy ecosystems in providing sustainable livelihoods and the importance of environmental governance in empowering the poor. UNEP's Inclusive Wealth Accounts initiative, implemented in partnership with IHDP-UNU, published the second inclusive wealth report. The report gives an overview of the evolution of some categories of natural capital for some countries. It indicated that over the period assessed, natural resources per capita declined by 33% in South Africa. If measured by GDP, the economies South Africa grew by 24% between 1992 and 2012. However, when South Africa's performance is assessed in terms of inclusive wealth it decreased by 1% UNU-IHDP and UNEP (2012, 2014).

The interactions between environmental change and human societies have a long, complex history spanning many millennia, but these have changed fundamentally in the last century. Ocean, coastal, climate and terrestrial ecosystems have degraded significantly. Global fisheries have been either fully exploited, overfished or collapsed. Coastal zones have been significantly altered and tropical rainforest and woodland have been lost. Especially over the last 60 years, humans have rapidly and extensively changed ecosystems mainly to meet rapidly growing demands for food, freshwater, timber, fiber and fuel (MA 2005). This trend is mainly driven by expanding flows of goods, services, capital, people, technologies, information, ideas and labor taking place in an increasingly globalized, urbanized and indus-trialized world.

The Millennium Ecosystem Assessment (MA) provides an elaborate concep-tual framework using the common denominator ecosystem services to describe all goods and services provided by natural capital. The MA defines ecosystem services as "the benefits that people obtain from ecosystems". Ecosystem services influence human well-being, and thus represent a value for society. The concept of ecosystem services is a strong tool to translate unnoticed benefits of nature into aspects of human well-being, which can be taken into account in decision making on proposed projects, programs, plans or policies.

Extrapolations for a range of indicators suggest that based on current trends, human pressure on ecosystems continues to increase, and that the degradation of ecosystem services could grow significantly worse during the first half of this century. This creates a barrier to achieving the Sustainable Development Goals, particularly those aimed at reducing hunger and poverty, improving human health and ensuring a sustainable supply of energy, food and clean water. How ecosys-tems are managed locally to globally would significantly impact the way in which water, food security and other ecosystem services are affected.

Based on the available evidence, public awareness of biodiversity has increased in both the developed and developing countries. Integration of values into planning process has been achieved in some countries. Current responses are not sufficient relative to pressures; therefore they may not overcome the growing impacts of the drivers of losses. However, governments still provide subsidies harmful to the environment. Policies and plans are needed for the removal, phasing out or reform of harmful subsidies. In addition to their adverse environmental effects, subsidies also have significant fiscal burden on the national budget. For instance, fossil fuel subsidies cost governments about US$410 billion in 2010. Removing envi-ronmentally harmful subsidies could boost the global economy by around 0.3%. In addition to the potential fiscal benefits, the removal of subsidies in developing and emerging economies might bring additional environmental benefits (OECD, 2006). Moreover, energy subsidies perform poorly as a means of supporting the incomes of poor social groups, considering that only 7% of the subsidies go to the bottom quintile in developing countries. By reforming fossil fuel subsidies, coun-tries can redirect those subsidies to support low-income households. However, the removal of subsidies alone is not likely to be sufficient to ensure a sustainable policy; taxing "bad behavior" is also needed to direct the incentives to greener investments. Environmental taxes have proven to generally be more effective in addressing environmental externalities than other more indirect taxes. Environ-mental taxes are cost effective tools in shifting technologies since they equate incremental costs of reducing emissions across firms, and sectors (OECD). The findings indicate that an emission tax is always welfare dominant over a subsidy on consumer purchases of the clean product because of its contribution to a reduc-tion in environmental damage. It does this by both inducing firms to improve the environmental qualities of their products and by constraining consumer usage of these products (Matsukawa, 2012).

There is a need for further compilation of environmental statistics and building environmental-economic accounts, including developing and maintaining national accounts of ecosystem services. Developing policies to address habitat loss and degradation, incentives and effective protected area networks are some key interventions needed to tackle with the fragmentation and degradation of habitats. However, it is important to diagnose the problems and potentialities of linked ecosystems and habitats, which requires serious study of complex, multivariable, nonlinear, cross-scale and changing systems and link these attributes with larger socioeconomic, political and ecological settings in which they are embedded (Ostrom, 2007). The most important policy decisions affecting ecosystems are often made in policy arenas other than those related to protecting ecosystems. For instance, development and diffusion of technologies designed to increase the efficiency of resource use or reduce the impacts of drivers such as climate change and nutrient loading will enhance provision of ecosystem services. For sustaining well-being, there is a need for factoring natural capital in development decisions.

Mainstreaming natural capital

Nature has been ill-served by 20th century economics. When asked, economists acknowledge nature's existence, but most would appear to deny that she is worth much. If ecologists worry about the contemporary nexus between population size (and growth), the standard of living and the natural environment, economists point to the accumulation of capital and technological progress and say Malthus got it wrong.

In a broad sense, mainstreaming of natural capital into development policy is designed to help decision makers at all levels (governments, businesses, multilevel development banks, individuals, etc.) integrate ecosystems and their services into their decision making (e.g. see Maes et al., 2013; Cowling et al., 2008). It follows that mainstreaming requires decision makers to consider both the positive and negative impacts of their choices on ecosystems and their services during the design and implementation of policies. Mainstreaming delivers on this objective by providing information and tools that can facilitate the connection of healthy ecosystems and sustainable well-being.

Mainstreaming facilitates policy to recognize the links between nature and development, considering the environmental and economic trade-offs associated with development measures and incorporating ES-related opportunities and risks into the implementation of strategies. This would require the integration of natural capital with other forms of capital into development policy which will eventually help assessment of the sustainability and effectiveness of development options. Mainstreaming natural capital into development decisions recognizes the missing link between economic and environmental policies.

An economic approach to ecosystem management is essential to analyse and assess the dynamics of human well-being and constituents of ecosystems, which have been elucidated by the Millennium Ecosystem Assessment (MA). The insights gathered from environmental economic analyses are critical for decision makers confronted with resource constraints and conflicting choices while designing and implementing development policies. Trade-offs between ecosystem services and human well-being can be better reconciled by adopting a quantitative and credible approach. The development of the ecosystem services paradigm has enhanced our understanding of the fact that natural capital is an indispensable form of capital asset along with produced and human capital. As Dasgupta argues, ecosystems are capital assets. Like man-made capital assets (e.g. buildings, and machinery), ecosystems depreciate if they are misused or are overused. But they differ from reproducible capital assets in three ways: (1) depreciation of natural capital is frequently irreversible (or at best the systems take a long time to recover); (2) except in a very limited sense, it isn't possible to replace a depleted or degraded ecosystem by a new one, and (3) ecosystems can collapse abruptly, without much prior warning (Dasgupta, 2008).

Arrow and Fisher (1974) first examined the public policy implications of the environmental costs of some economic activities. An accurate determination of the comparative economics of preservation versus development of a pristine natural area is often rendered difficult because of the uncertainty involved in the estimates of environmental costs (Arrow, and Fishe 1974). The value of a provisioning service is often embedded in measures of gross domestic product (GDP), but often ignored. For example, the water used in agricultural production contributes to economic value, but is seldom recognized as a separate component of GDP. Regulating services can also have direct values that are directly or embedded in GDP, but will usually provide benefits that have value, but not the type that ends up in GDP (Managi and Kumar, 2018).

Some studies have shown that macroeconomic policies, which are especially called as structural adjustment policies, might have a significant impact on ecosystem services. The devaluation, liberalization and interest rate adjustment policies might create different incentives on producers or the consumers which might cause adverse effects on land (soil/terrain) and water use and biodiversity. In Cameroon, for example, after the devaluation decision, producers prices increased which caused an expansion of unsustainable cultivation on to marginal land. Intensive methods of cultivation were penalized by higher costs of inputs. Some studies show that as a result of devaluation there has been a switch to the production of high value horticultural crops with mixed environmental impacts. In several countries liberalization has led to increased deforestation. Evidence suggests that increased deforestation has been one of the main consequences of land-use changes brought about by higher interest rates and a shift of price incentives towards tradable commodities. An increase in the price of commercial fuel as a result of reduced public subsidy increased the use of fuel wood, leading to increased deforestation in Cameroon. In Ghana there has been a major growth of logging since 1984 as a result of deregulation and increased price incentives (Ahmed and Lipton, 1997).

Green economy policies which focus on both environmental improvements and economic development offer practical and applicable tools with emphasis on reforming economic instruments for mobilizing financing for green investment, managing environmentally harmful products, greening SMEs, reforming environmental harmful subsidies and developing a set of indicators for measuring progress towards sustainable development. Green growth policies will also facilitate the country to achieve recently adopted Sustainable Development Goals.

Natural capital plays an important role in poverty reduction; however, positive poverty reduction and ecosystem services outcomes cannot be taken for granted. Mainstreaming ecosystem services needs careful consideration because many of the opportunities identified can reduce poverty but may have the opposite effect if poorly managed or implemented (Kok et al., 2010). Nature is a one of the key capital forms of economy and should not be seen as the sink for residuals. The poor are the most seriously impacted social group by ecosystem losses. Five hundred forty million people are engaged in farming, animal husbandry, informal forestry and fisheries. Ecosystem services add only 7.3% to conventional GDP; however, ecosystem services add 57% to the "GDP of the Poor". Replacement of the losses of ecosystem services is beyond the capacity of the poor.

The majority of ecosystem services are not marketed, and it is often difficult to determine how changes in ecosystem structure, functions, and processes influence the quantities and qualities of ecosystem service flows to people (Barbier, 2011). Valuation of ecosystem services may also lead to the use of some market mechanisms. The heart of the Coasian approach lies in voluntary negotiated agreements among the affected parties. Such agreements, Coase argues, would result in efficient resource allocation, so long as the property rights are well defined and the costs of negotiation or transaction are negligible (Coase, 1960). The main allocative function of property rights is the provision of incentives to ensure a "greater internalization of externalities." Property rights emerge from the desires of the interacting parties to adjust to new benefit-cost possibilities (Demsetz, 1967). PES schemes should focus on cost-effectiveness and best practice for positive livelihood impacts. PES schemes should be transparent, and provide additional services with conditional payments to voluntary providers (Luca, 2012).

There is a growing interest in mainstreaming natural capital into development policy, and many international initiatives have been complementing this expanding interest. The Poverty-Environment Initiative (PEI) of the UNDP and the UNEP is a global programme that supports country-led efforts to mainstream poverty-environment linkages into national development and sub-national development planning, from policy making to budgeting, implementation and monitoring. PEI has strengthened coordination at the national level by working through ministries of planning and finance and connecting those to the environment sector. This initiative created enabling conditions at policy and institutional levels across sectors. This initiative also catalysed the national funding for country programmes.

UNEP (2011) defines a green economy as one that results in "improved human well-being and social equity, while significantly reducing environmental risks and ecological scarcities". Countries have recognized a green economy as a vehicle for achieving sustainable development and poverty eradication, and there is a growing demand from countries to implement green economy strategies. In response

to this demand, the UNEP, ILO, UNDP, UNIDO and UNITAR created the Partnership for Action on Green Economy (PAGE) initiative. PAGE is offering a full
range of services and tools which helps countries to implement their national green
economy plans.

Four lessons are highlighted: (1) Information on ES is considered useful by
many practitioners, but the type, production, and communication of ES information need to be adapted to the specific context of a planning case; (2) A broad
range of approaches is available for integrating the ES concept in (participatory)
planning with different and complementary contributions to decision-support; (3)
Effectively integrating ES in planning requires careful scoping of the context,
objectives, and capacities; (4) Integrating ES in planning can effectively support
the co-production of relevant knowledge and the collaboration of diverse actors
(Albert, 2014).

Besides the findings of these global and regional initiatives, more information is
needed that considers the full ensemble of processes and feedbacks, for a range of
biophysical and social systems, to better understand and manage the dynamics of
natural and human systems. Such information will expand the capacity to address
fundamental questions about complex social-ecological systems (Carpenter et al.,
2009).

In this backdrop this book introduces new cases and research results with a
specific context and provides examples how different mainstreaming tools and
methodologies could effectively be integrated into development planning, implementation and evaluation.

Chapter 2 demonstrates that there is a growing interest in the potential of strategic environmental assessment (SEA) to mainstream ecosystem services (ES)
concerns in decision making. Experiences in this field have begun to emerge, highlighting the need for comprehensive guidance. This chapter addresses this need by
proposing a conceptual approach to integrate ES effectively in SEA. The approach
is structured in the following four stages, each comprising specific tasks: establish
the ES context; determine and assess priority ES; identify alternatives and assess
impacts on ES; follow up on ES. The first stage includes the identification and
mapping of ES and beneficiaries for the region affected by the strategic action and
the identification of links with other strategic actions. The second stage aims at
generating detailed information on a limited set of priority ES, which are considered
relevant for shaping and informing the development of the strategic action. This
requires determining the priority ES, reviewing existing regulations concerning
these services, and assessing their baseline conditions and trends. In the third stage,
possible alternatives to enhance ES, or at least to minimise negative effects on them,
are identified and their impacts assessed. Finally, during the fourth stage, the effects
on ES are monitored and managed and the overall quality of the SEA process is
tested. The chapter oncludes by discussing how the stages and their tasks require
feedback and interactions and how they can contribute to achieve a better inclusion
of concerns about ES (and their beneficiaries) into strategic decisions.

Chapter 3 picks up a case from South Africa. South Africa has a long history
of ecosystem service assessments, with many examples of how to engage in

multi-stakeholder processes that can facilitate the co-production and exchange of knowledge. The chapter reflects on a handful of these mainstreaming activities in South Africa and discusses some lessons learnt on the opportunities and challenges of integrating information on ecosystem services into decision making processes at different scales and in different contexts. The approach here centers on three iterative phases: project co-design, knowledge and output co-production and co-implementation activities that can assist with learning and adaptation. The authors conclude that transdisciplinary approaches that embed activities in a targeted implementation context and easily respond to windows of opportunity are more likely to result in mainstreaming success.

Chapter 4 demonstrates the application of a dynamic and general equilibrium model to measure the shadow rental values of water in the Aral Sea. They analyse policy options of calculating the asset value of land and water, then they evaluate how the water trading in the region affects the condition of the Aral Sea. Finally, they analyse how the canal efficiency might influence the asset value of water and land in the region which has strong implications for sustainability of Aral Sea.

It introduces a tool for mainstreaming ecosystem service valuation that helps in understanding the impact of policy on natural asset wealth, and illustrates how to use the tool and interpret the results. The focus of the chapter is two regions in Kazakhstan: South Kazakhstan and Kyzylorda. The authors analyse three policy options: the status quo, trading water use rights and improving irrigation efficiency. The major contribution of the chapter is to demonstrate how land and water asset (stock) values can be used to evaluate the impact of policy. Natural resource stock values provide what appears to be a natural index to use to compare the impact of policy on natural resource (or ecosystem) wealth. The model introduced in this chapter examines and predicts: (1) current and future levels of gross domestic product (GDP) across seven Kazakh subsectors, (2) the contribution of water and land to agricultural value-added (loosely viewed as GDP) along the Syr Darya, and agricultural value-added in the rest of Kazakhstan and (3) the stock (or wealth) value of land and water in Kazakhstan. The authors suggest that trading water use rights could increase the wealth/well-being of those controlling the use rights of land and water by 9%. Irrigation improvements, however, yield smaller gains (less than 1%). The study also illustrates how policy can impact specific natural asset values in different ways. For example, trading water use rights leads to a decrease in the stock (wealth) value of water, but an increase in the stock value of land. Improving irrigation efficiency tends to increase the wealth values of both land and water.

Models play an important role to identify the optimal solution by assessing the outcomes of the different options. Yet, institutions and individuals play important roles in making the change to a sustainable development path. However, a behavioral change at both personal and institutional scale cannot be successful unless the persons or institutions internalize the reform. This can only be achieved if key stakeholders are part of this change process.

Chapter 5 suggests that capturing tourism and recreational benefits from various ecosystems can help their preservation agenda. It may also play an important role in the conservation and sustainable management of natural ecosystems and cultural heritage sites. Local, domestic and international visitors, however, often differ in their cultural preferences as well as in the spatial distribution of their visitation patterns. The investigation of such differences is usually achieved through surveys which, however, are time- and resource-intensive and often impractical at large scales. In this chapter the spatial distribution of cultural ecosystem services and historical heritage tourism and recreation has been investigated. These benefits are enjoyed by local, domestic and international visitors to the Usumacinta floodplain, a 25,000 km² coastal region with one of the highest biological and cultural diversities in Mexico. The metadata of 8,245 geotagged photographs taken within the boundaries of the Usumacinta floodplain and uploaded by 499 individual users to the popular photo-sharing website Flickr has been analysed. A GIS-based kernel density analysis of the photographs' spatial distribution shows that culturally tagged photographs taken by international visitors tend to be concentrated in correspondence to famous archaeological sites such as the Mayan ruins of Palenque, while those from local and domestic visitors are more widespread within the floodplain. The chapter advocates based on the findings that that the improved understanding of spatial patterns and different cultural benefits accrued to various segments of the population that can be obtained through the monitoring and analysis of their social media activity, as demonstrated in this study in application to the Usumacinta floodplain, has important and often overlooked implications for the sustainable management of natural resources and cultural heritage sites.

Chapter 6 discusses how mainstreaming tools strengthened the relationships across the different local stakeholders and built trust among scientists, state organizations and the local communities. Authors introduce a case from San Pedro de Atacama, Chile. As a mainstreaming tool, the researchers developed a Water Balance Model which was used to simulate the hydrological balance in the Atacama plateau and the underlying inter-connections between ecotourism, recreation, the condition of this desert area, and its potential for water provisioning. Management of water provisioning is important not only to recreationists and the local population but also important in supporting local biodiversity, including birds such as the flamingo, which are tourist attractions. The model is at the core of a decision support system that is used for municipality land use management policy and for regional biodiversity conservation. The authors highlight the importance of making the data on water demand, tourism flows and biodiversity available for the community. This trust-building capital facilitated the mainstreaming, scaling-up and applicability of these tools to other areas in Chile.

Many challenges exist for mainstreaming ecosystem service at a local level; however there are also strong opportunities in the multi-sectoral planning processes driving development and in how the concept of ecosystem services is framed and aligned with development priorities, especially those relating to disaster risk reduction (Sitas et al., 2014).

Chapter 7 discusses the mainstreaming of biodiversity and ecosystem services into development planning in the context of Small Island Developing States. First, this chapter addresses the integration of biodiversity and ecosystem services considerations into land use and spatial planning. In this context, GIS-based maps for spatial planning have also been made available to the ministry for their use – these include (1) pollination maps for the Nariva Swamp, Ramsar site; (2) maps on sediment retention and water purification for the Northern Range in Trinidad and (3) coastal vulnerability maps for southwest Tobago. Second, this chapter introduces the methodology that was developed and adapted for Capital and Ecosystem Services Accounting for Trinidad and Tobago. On the basis of this, demonstration ecosystem services accounts have been produced for water, carbon, biodiversity and land.

Finally, Chapter 7 refers to the exploration of the potential for developing Payment for Ecosystem Services (PES) pilot. A roadmap has been put together and is ready for piloting and implementation.

Mobilizing political support, required changes in legislation, budgetary concerns, capacity building and organizational management are the issues that need to be addressed for successful integration of ecosystem service framework. Among these tasks, having a common vision, i.e. linking and merging environmental goals with development goals, is one of the initial steps for successful implementation. High level involvement is as important as popular involvement. The chapter comes out with the strategies for mainstreaming that delivered communication and outreach and co-production of knowledge, supported the integration of ecosystem services into national policy and dialogue, including the production of national ecosystem service maps, promoted public-private cooperation for ecosystem management, developed ecosystem-service models to inform investments and established a framework for investment in ecological infrastructure that in turn has been adopted in national policy and planning.

Chapter 8 provides insights into how ecosystem services concept can be mainstreamed to nationwide strategies. The authors provide a case study from the Cape Ca Mau National Park/Biosphere Reserve and the surrounding Mekong Delta, in the Ca Mau Province – covering 12% of the country and comprising one of the largest remaining contiguous mangrove forests in the country. The local mainstreaming objective is highlighted as supporting the Ca Mau Division of Natural Resources and Environment and the Ca Mau National Park management to integrate ecosystem services into land use planning by developing ecosystem service models to assess carbon storage and coastal vulnerability of Ca Mau ecosystems and by conducting an ecosystem service valuation study which demonstrates that the economic value of the coastal protection service provided by mangroves in Ca Mau averages US$2,600 per hectare per year, which is 25 times more than timber market value of mangroves. This chapter presents the results of applying valuation of ES and mapping tools to support mainstreaming efforts at different levels which contribute to conserve the important mangrove ecosystem.

Chapter 9 also demonstrates the use of economic tools to assess whether there is an economic rationale for the reduction of deforestation in the United Republic of Tanzania. This country hosts 48 million hectares (ha) of forest, which is 51% of the country's area. At the same time, annual deforestation is estimated by the National Forest Monitoring and Assessment (NAFORMA, 2014) at 372,816 ha between 1995 and 2010. In this context, this chapter presents and discusses the results of a cost-benefit analysis with the objective to estimate the present value of net economic impact from deforestation to the Tanzanian economy over the next 20 years (2013–2033). According to the estimation results, the net economic losses from deforestation to the Tanzanian economy over the next 20 years are valued to US$171 million. This economic value is, however, interpreted as a lower bound estimate as the economic analysis included only timber resources, a forest provisioning ecosystem service, which is directly captured by the system of national accounts and can therefore be reflected in the gross domestic product (GDP). Additional analysis also highlights that investments in the forestry sector lead to comparatively higher income for rural populations than the same investments in the sectors of agriculture and wood, paper and printing. Hence, investments in the forestry sector can potentially also be beneficial from the perspective of poverty alleviation.

These findings highlight that it is economically interesting for the United Republic of Tanzania to invest in conserving its forests, and therefore present a case for the government to tackle the direct and underlying drivers of deforestation and transition, moving towards an economic model that stimulates sustainable use and conservation of forest ecosystems by implementing REDD+. In that sense, this report provides further rationale for efforts to accelerate the implementation of the REDD+ National Strategy and Action Plan.

Chapter 10 provides an assessment of the "readiness" in Vietnam to implement measures to support Vietnam's goals of achieving green and equitable growth. Based on a review of key national legislations, it is recommended that a Green Economy–System Dynamic Modelling (GE-SDM) approach using three indicators – GDP of the Poor, Decent Green Jobs, and Green GDP – is the most effective tool for Vietnam to meet the goals listed under its National Green Growth Strategy. SDM is a computer modelling system that allows policy makers to compare potential economic trajectories for given sectors over time. Indicators are used as inputs and the model features feedback loops and delays that mimic the complications of a real economy (Sterman, 2000). In the case of a green economy, planners can derive understanding and policy insights from the visualized impacts on the economy, environment and poverty that are demonstrated by the three indicators in a System Dynamic Model. According to primary data and model results the results, Vietnam is considerably advanced in its legislation and National Green Growth Strategy targets, but the country would benefit from an approach that allows the government to evaluate the success of its initiatives thus far in achieving desired goals. In particular, a GE-SDM approach, encompassing appropriate green economy indicators, will support Vietnam by providing an effective set of green economy metrics which identify policy gaps to be improved. In this context, if Vietnam selects a GE-SDM approach, it is therefore recommended that a national level analysis is undertaken to estimate Green GDP and Decent Green Jobs and a pilot province is selected to test the implementation of a GDP of the Poor indicator.

Chapter 11 addresses the issue of discounting while dealing with natural capital. Rate of discount helps the intertemporal evaluation of the impact on natural capital. It also discusses the importance of discount rates in determining the viability of different kinds of investment in natural capital. Individual preferences are not of normative significance for government decisions, and time and risk preferences relevant for government decisions should be developed as a matter of national policy (Arrow and Lind, 1970). However, Baumol argues that it is not possible to determine precisely the social rate of discount, given the complexity of our institutional arrangements (Baumol, 1968). The discount rate is used to compare economic effects that occur at different points in time. Societies as whole and individuals within them separately prefer, for different reasons, to have something now, rather than to have the same thing in the future.

High rates militate against investments that produce benefits after a long gestation period or cases where the benefits are spread out over a very long period. Often benefits are ignored when they do not generate monetary flows.

Since natural capital is a public good, or has a large element of its services that are a public good, the discount rate for valuing natural capital should be the social rate. The theoretical discussion on social rates provides a basis for choosing the value to be used. The author suggests that the social preferences are best represented by declining rates, so that services received in the short run are discounted at a higher rate than services received in the long run. Such a structure captures better than a constant discount rate social preferences on costs and benefits at different points in time. Declining rates also capture the impact of uncertainty in comparing flows at different points in time. Some governments (United Kingdom, France, partly also the United States) have adopted declining discount rates to value costs and benefits in long term projects. The practical literature on valuing natural capital has dealt quite cursorily with the choice of the discount rate. The author provides examples of different initiatives and institutions which choose a social discount rate varying between 1% and 5%. Time horizon limitation is also another practice in valuation of natural capital. Time limitation is justified on the grounds that at typical discount rates the benefits after 25 years play a small part in the total value and that estimates of other components of future value beyond that period have very high uncertainties.

Conclusions

Natural capital is critically important component of societal consumption and production. It is just like any other capital which needs investment in terms of maintenance costs or alternatively investment through restoration. Maintenance of natural capital would enable it to provide much needed services to society. It is important that a scientific assessment, credible accounting and robust valuation are done at local, national, regional and global scales. Once their measurement and valuation are done, integration of natural capital into the design of development – especially poverty alleviation goals, fiscal reforms, trade and investment design, implementation and evaluation of projects – would become easier. Natural capital has special significance for poverty alleviation as their nexus is deep and all pervasive. This has been demonstrated by programmes like the poverty-environment initiative and ecosystem services for poverty alleviation. UN environment implemented programmes like the Project for Ecosystem Services (ProEcoServ) in four countries – Chile, South Africa, Trinidad and Tobago and Vietnam – clearly show that that the trade-off amongst natural and other types of capital is also possible at different scales and for different ecosystem services.

It is also evident that additional scientific knowledge would be needed for ecosystem service production functions and trade-offs.

We hope that the practitioners as well as the researchers would find the contributions in this volume useful and catalytic enough for further exploration of the issues in natural capital so that society can find the investment value in the restoration of natural capital instead of costs.

References

Ahmed, I., and Lipton, M. (1997). Impact of Structural Adjustment on Sustainable Rural Livelihoods: A Review of the Literature. IDS Working Papers, 62.

Albert, C., Aronson, J., et al. (2014). "Integrating ecosystem services in landscape planning: Requirements, approaches, and impacts." *Landscape Ecology* 29(8): 1277–1285.

Arrow, K. J., and Fisher, A. C. (1974). "Environmental preservation, uncertainty, and irreversibility." *The Quarterly Journal of Economics* 88(2): 312–319.

Arrow, K. J., and Lind, R. C. (1970). "Uncertainty and the evaluation of public investment decisions." *The American Economic Review* 60(3): 364–378.

Assessment, M. E. (2005). "Millennium ecosystem assessment." In *Ecosystems and Human Well-Being: Synthesis*. Washington, DC: Island Press.

Barbier, E. (2011). *Capitalizing on Nature: Ecosystems as Natural Assets*. Cambridge, UK; New York: Cambridge University Press.

Baumol, W. J. (1968). "On the social rate of discount." *The American Economic Review* 58(4): 788–802.

Brundtland, G. (1987). Report of the World Commission on Environment and Development: Our Common Future. United Nations General Assembly document A/42/427.

Carpenter, S. R., Mooney, H. A., et al. (2009). "Science for managing ecosystem services: Beyond the millennium ecosystem assessment." *Proceedings of the National Academy of Sciences* 106(5): 1305–1312.

Coase, R. H. (1960). "The problem of social cost." *The Journal of Law and Economics* 3: 1–44.

Cowling, R. M., Egoh, B., Knight, A. T., O'Farrell, P. J., Reyers, B., Rouget, M., Roux, D. J., Welz, A., and Wilhelm-Rechman, A. (2008). "An operational model for mainstreaming ecosystem services for implementation." *Proceedings of the National Academy of Sciences of the United States of America* 105: 9483–9488.

Dasgupta, P. (2008). "Nature in economics." *Environmental and Resource Economics* 39(1): 1–7.

Demsetz, H. (1967). "Toward a theory of property rights." *The American Economic Review* 57(2): 347–359.

Kok, M. T. J., Tyler, S., et al. (2010). "Prospects for mainstreaming ecosystem goods and services in international policies." *Biodiversity* 11(1–2): 49–54.

Luca, T. (2012). "Redefining payments for environmental services." *Ecological Economics* 73(1): 29–36.

Maes, J., Teller, A., Erhard, M., et al. (2013). *Mapping and Assessment of Ecosystems and Their Services: An Analytical Framework for Ecosystem Assessments under Action 5 of the EU Biodiversity Strategy to 2020*. Luxembourg: Publications Office of the European Union.

Managi, S., and Kumar, P. (2018). *Inclusive Wealth Report*. Abingdon, UK: Routledge.

Matsukawa, I. (2012). "The welfare effects of environmental taxation on a green market where consumers emit a pollutant." *Environmental and Resource Economics* 52(1): 87–107.

NAFORMA (2014). *The National Forest Resources Monitoring and Assessment 2009–2014*. Tanzania Forest Service Agency, United Republic of Tanzania, Dar es Salaam.

OECD Sustainable Development Studies. (2006). *Subsidy Reform and Sustainable Development: Economic, Environmental, and Social Aspects*. Paris: OECD Publications.

Ostrom, E. (2007). "A diagnostic approach for going beyond panaceas." *Proceedings of the National Academy of Sciences* 104(39): 15181.

Polasky, S., et al. (2019). "Role of Economics in analyzing the environment and sustainable development." *Proceedings of the National Academy of Sciences of the United States of America* 116(2), March 19.

Sitas, N., Prozesky, H.E., Esler, K.J., and Reyers, B. (2014) "Opportunities and challenges for mainstreaming ecosystem services in development planning: Perspectives from a landscape level." *Landscape Ecology* 29: 1315.

Sterman, J. (2000) *Business Dynamics: Systems Thinking and Modeling for a Complex World*. Boston: Irwin/McGraw-Hill.

UNEP (2011) *Towards a Green Economy: Pathways to Sustainable Development and Poverty Eradication – A Synthesis for Policy Makers*. www.unep.org/greeneconomy

UNU-IHDP and UNEP. (2012, 2014). *Inclusive Wealth Report 2012: Measuring Progress toward Sustainability: Summary for Decision-Makers*. Bonn: UHU-IHDP.

2 Strategic Environmental Assessment as a tool to mainstream ecosystem services in planning

Davide Geneletti

Introduction

What is Strategic Environmental Assessment?

Strategic Environmental Assessment (SEA) refers to a "range of analytical and participatory approaches that aim to integrate environmental considerations into policies, plans and programmes and evaluate the interlinkages with economic and social considerations" (OECD, 2006). SEA applies primarily to development-related initiatives promoted individually in sectors (e.g., transport, energy, water, tourism) or collectively in a geographical area (e.g., regional spatial or land use plan). The field of SEA has developed rapidly over the last 15 years, and SEA is undertaken, both formally and informally, in an increasing number of countries and international organizations (Sadler, 2011). Two EU-led initiatives promoted the application of SEA. Firstly, the European SEA Directive (2001/42/EC) requires an environmental assessment for certain plans and programmes at various levels (national, regional and local) that are likely to have significant effects on the environment. Secondly, the SEA Protocol to the Espoo Convention (UNECE Convention on Environmental Impact Assessments [EIA] in Transboundary Contexts), agreed on in 2003, encourages the use of SEA in the context of policies and legislation (OECD, 2006). Currently, several dozen countries around the world have either national legislative or other provisions for SEA, e.g. statutory instruments, cabinet and ministerial decisions, circulars and advice notes. Increasingly, developing countries are introducing legislation or regulations to undertake SEA or are applying SEA-type processes (Sadler, 2011).

Even though SEA approaches vary in different countries, for different sectors and for different levels of decision making, there is broad agreement on certain defining principles (Therivel, 2004):

- SEA is a tool for improving strategic actions. Hence, SEA should start early, and be undertaken as an integral part of the decision-making process. Decision makers should be involved in the SEA process to ensure that proper considerations is given to SEA findings.
- SEA should promote stakeholders' participation and ensure transparency in the decision-making process, including sensitivity to gender.
- SEA should focus on key environmental and sustainability concerns that are appropriate for the specific strategic action, considering the timescale

and resources of the decision-making process. A scoping stage is always important to sort out the key issues.

- SEA should include the analysis and comparison of possible options for the strategic action, and the identification of the most suitable one(s).
- SEA should aim at minimizing negative effects, enhancing positive ones, compensating for the loss of valuable features and benefits, and ensuring that irreversible damage are not caused. This requires predicting the effects of the strategic decision and comparing the likely future situation without the action (the baseline) against the situation with the action. It also requires evaluating the significance of the effects.

In short, a good-quality SEA process informs planners, decision makers and affected public on the sustainability of strategic decisions, facilitates the search for the best alternative and ensures a democratic decision-making process (IAIA, 2002).

There are several reasons why SEA is important, including (after Partidário, 2012; Sadler, 2011; Abaza et al., 2004):

- Promotes environmentally sound and sustainable development, shifting from a "do least harm" to "do most good" approach.
- Allows problems of environmental deterioration to be addressed at their "upstream source" in policy and plan-making processes, rather than mitigating their "downstream symptoms" at project level, extending the principles of Environmental Impact Assessment.
- Provides early warning of large-scale and cumulative effects, including those resulting from a number of smaller-scale projects.
- Facilitates identification and discussion of development options and provides guidelines to help development to follow sustainability trajectories.
- Encourages political willingness, stimulates changes to mentalities and creates a culture of strategic decision-making.

For these reasons, SEA is considered one of the most promising tools to integrate environmental concerns into strategic decision making, and more broadly to help facing development challenges (World Bank, 2009).

How is SEA applied?

SEA must be flexible and able to adapt to the planning and policy-making context (including legal, institutions, procedural and political factors), which may be very different among countries, decision tiers (national, regional, etc.) and sectors (land use, agriculture, water, energy, etc). Additionally, the specific circumstances of the strategic action under consideration (in terms of content, level of definition, availability of data, timing, consultation with stakeholders, etc.) will determine the way in which SEA is undertaken. Therefore, SEA can be applied in various ways to suit particular needs. For example, some SEAs (OECD, 2006):

- Are "stand-alone" processes running parallel to core planning/policy processes, while others are integrated into them.

- May focus on environmental impacts, while others integrate all three dimensions of sustainability: environment, social and economic.
- May engage a broad range of stakeholders or be limited to expert policy analysts.
- May consist of a quick analysis in a short time frame while others require detailed analysis over a long period.
- Can be a finite, output-based activity (e.g. a report), or a more continuous process that is integrated within decision making, focused on outcomes, and that strengthens institutional capacity.

In the light of this, a number of methodological approaches have been proposed over the years to tailor SEA to different decision-making contexts and to show the broad range of possible SEA forms (e.g., some are more focused on the assessment of impacts, others on institutions and strategic thinking). Even though SEA cannot be represented by a standard sequence of activities, the SEA principles described earlier allow for the identification of a number of typical stages through which SEA can feed decision making. Figure 2.1 presents the SEA

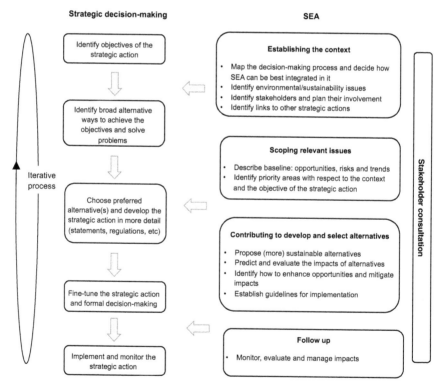

Figure 2.1 Strategic decision making and SEA stages.

Source: Geneletti (2015) (building on Therivel, 2004)

stages associated with the broad stages of strategic decision making. In reality, decision making seldom follows this idealized model; hence SEA needs to operate under different conditions and be flexible. More details on the activities to be performed under each stage and on possible variants, can be found in OECD (2006), Fischer (2007), Ahmed and Sánchez-Triana (2008), Therivel (2004), and Partidário (2012).

Ecosystem services in SEA

The MA (2005), as well as many other studies that followed, highlighted the importance of integrating ecosystem services (ES) into strategic decision-making, so that the effects of certain development options on ecosystems, and the services they provide, can be considered at the earliest appropriate stage. SEA is potentially a very suitable tool to achieve this because – as described in the previous sections – it focuses on strategic actions, promotes stakeholder engagement and helps decision makers face broad development challenges. Additionally, the content of SEA is increasingly extending beyond the biophysical environment to include also other issues (social, health, economic) associated with human well-being. Typically, policy-makers are interested in knowing how the costs and benefits are shared by the stakeholders of society in execution of policies. The concept of ESs is a viable tool for impact assessments, as it provides a means to translate unattended and unintended consequence of policy implementation on human well-being (Kumar et al., 2013).

The use of the ES concept in SEA also offers the advantage of presenting a more holistic and integrated consideration of the socio-ecological system and an effective framing of the (natural) environment in terms of communicating with and influencing stakeholders and decision makers (Baker et al., 2013). ES can be appropriate indicators to weight impacts of developmental policies, programmes and plans on the state of environment, which is the basis for providing natural capital. Using ES as indicators could help ensuring that appropriate considerations are given to the implications of environmental impacts when introducing new developments. Table 2.1 summarises the possible contribution of ES to good-quality SEA.

There is a growing interest in the potential of SEA, and impact assessment in general, to mainstream ES concerns in decision making, as shown by recent scientific publications (Mandle et al., 2015; Geneletti, 2016a; Geneletti et al., 2015; Geneletti, 2013a; Karjalainen et al., 2013) and reviews of practices (Rosa and Sánchez, 2015; Honrado et al., 2013). Experiences in this field have begun to emerge in the last few years (Partidário and Gomes, 2013), showing the need for comprehensive guidance (Helming et al., 2013). This chapter addresses this need by proposing a methodological approach structured into a set of key stages and tasks to integrate ES effectively in SEA, focusing particularly on the planning level of decision-making.

Table 2.1 Examples of contributions of ES information to the quality of SEA

Characteristics of good-quality SEA (IAIA, 2002)	Contribution of ES information
Integrated	ES inherently address the interrelationships between biophysical and socio-economic aspects. The analysis of ES-related scale issues facilitates the interaction with relevant plans and policies at different decision-making tiers.
Sustainability-led	ES approaches explicitly link changes in ecosystems and biodiversity with effects on human well-being. Hence, ES-inclusive SEA processes extend beyond the assessment of biophysical and environmental factors only and promote plans that are more sustainable.
Focused	ES approaches offer a key to read the most important interactions between human society and the environment, identifying issues that are important for the specific decision-making context.
Accountable	Analysis of expected future trends in ES under different scenario conditions can be used to document how sustainability issues were taken into account and to justify planning choices.
Participative	Information on ES by definition requires the identification of beneficiaries and stakeholders (including by gender), paving the way to more participative SEA processes.
Iterative	The analysis of ES can be included, in different forms, throughout the whole process, so as to provide information on the expected impacts of plan's choices during the different "decision windows" of the planning/policy-making process.

Source: Geneletti (2011)

Building on the general SEA stages presented in Figure 2.1, the approach is structured in the following four stages, each comprising specific tasks (Figure 2.2):

- Establish the ES context (Stage 1)
- Determine and assess priority ES (Stage 2)
- Identify alternatives and assess impacts on ES (Stage 3)
- Follow up on ES (Stage 4)

Each successive stage in the proposed approach builds on previous work, but the sequence is not intended to be followed strictly. SEA is an iterative process, and many tasks may take place in parallel or in an order different from that presented here, according to the particular needs of the specific case, as explained in the coming sections. The overall purpose of the approach is to ensure that all relevant information of ES is collected, processed and used to support decision making. Stakeholder consultation is a vital component of SEA, and it is relevant in all the stages, as shown by Figure 2.2. Timely and well-planned consultation programmes facilitate the development of a shared vision of problems and objectives, contributing to the successful design, implementation and management of plans and policies (Slootweg et al., 2006; Abaza et al., 2004).

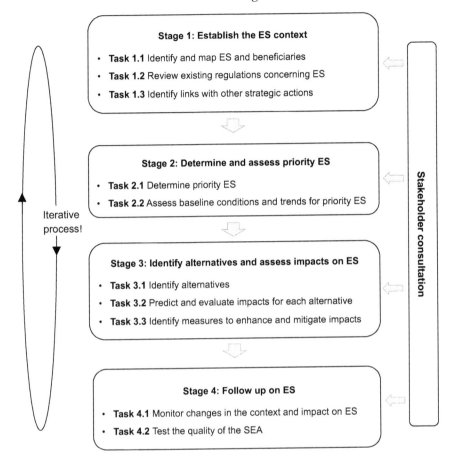

Figure 2.2 Stages and tasks to integrate ES in SEA.
Source: modified after Geneletti (2015a)

Stage 1: establish the ES context

In this first stage, SEA needs to provide an understanding of the context within which the plan will be developed and implemented. This requires identifying and mapping ES and beneficiaries for the region that will be affected by the plan (Task 1.1), reviewing existing regulations concerning these services (Task 1.2) and identifying links with other existing or foreseen strategic actions (Task 1.3).

Task 1.1: identify and map ES and beneficiaries

In order to incorporate information on ES into SEA, a general understanding of how ecosystem services are produced and used in the planning region needs to be achieved from the very beginning of the process. This can be obtained by: (1) identifying the main ecosystem types occurring in the study area, (2) determining the services produced by these ecosystems, and (3) describing the beneficiaries of such services (disaggregated by gender and other sensitive groups, if possible) and the contribution provided to their well-being (e.g., in terms of health, material assets, security). Alternatively, one may start by identifying the key elements of well-being for the region's inhabitants, whether or not they are shaped by ES. Then, the ecosystem goods and services that matter the most for those elements should be identified, and traced back to the ecosystems that supply them.

This task requires building a conceptual framework to link socio-economic systems with ecosystems, via the flow of ES. Many frameworks have been proposed for this purpose, including the MA (2005), TEEB (TEEB, 2011) and IPBES (IPBES, 2013) frameworks, the ES cascade model (Haines-Young and Potschin, 2010) and the EU framework for ecosystem assessments (Maes et al., 2013). All these conceptual frameworks relate to one another to some extent, even though they introduce differences, for example in the description of the well-being components or in the definition of the relationships between ecosystems and the values provided to people. Practitioners can refer to these frameworks to identify the most suited to their specific SEA context. Figure 2.3 presents an example of application of the MA framework to inform the early stage of SEA for a municipal development plan. This analysis allows us to identify the main direct and indirect drivers of ecosystem change and their effects on ES and ultimately on human well-being.

Whenever possible, details should be added concerning the relevance of ES for the well-being of different groups of beneficiaries, with specific attention paid to

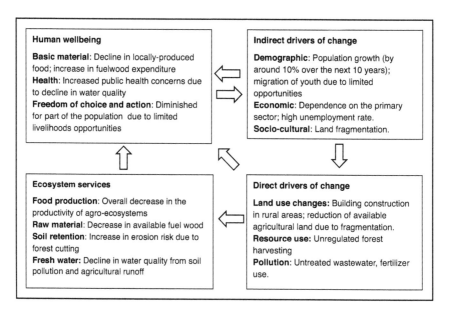

Figure 2.3 Analysis of the interaction between socio-economic and ecological systems in the municipality of Partesh/Partes (Kosovo) to support the SEA of the Municipal Development Plan.

Source: Geneletti (2016b)

the most vulnerable groups in terms of geographical location, as well as socio-economic conditions (e.g., by considering the level of dependence of different livelihoods on a given ES and the "substitutability" of that service). In addition, it is desirable to have also a (rough) geographical indication of where ES are produced and used.

Stakeholder consultation is essential for this task. Stakeholders' opinions can help to simplify the problem and get the essentials right. For example, participatory mapping approaches can be undertaken to gain a clearer view on what matters for people's well-being, and how this is associated to ecosystems and their services. Men and women often have different roles, albeit related, in the use and management of natural resources and ES. It is essential, therefore, to take these roles into consideration during the design of stakeholder consultation activities, as well as the subsequent SEA stages.

Task 1.2: review existing regulations concerning ES

ES often have some form of legal protection. Hence, a first and obvious step is to make sure that the plan is at least compliant with existing regulations and legal obligations. Of course, policy making should comply with legal obligations irrespective of whether an SEA is carried out or not. However, SEA makes sure that this is actually performed, so as to enhance the compatibility of plans and policies with the existing regulatory and strategic frameworks (see also next task). This task can be broken down into three activities. Firstly, identify all the existing regulations and legal obligations that set conditions for the use or protection of ES in the region. This activity is made difficult by the fact that regulations may contain "hidden", or anyway not explicit, references to ES. For example, an act concerning indigenous people may contain implications on how land need to be used and managed to ensure supply and fruition of ES (e.g., access to religious or cultural sites, food supplies, traditional medicines, etc.).

Examples of possible regulations to be reviewed include (modified after Slootweg et al., 2006):

- Provisioning services
  - Extractive reserves (forest, marine, fisheries)
  - Areas of high-quality soil
  - Areas of indigenous interest
  - Groundwater and surface water protection areas

- Regulating services
  - Urban and regional regulations on impervious surfaces
  - Flood storage areas
  - Regulations on forest and pasture for preventing hazards
  - Regulations on riversides
  - PES (Payment for ecosystem services) schemes

- Cultural services

  - Natural monuments, natural heritage sites and cultural heritage sites
  - Archaeological parks
  - Sacred sites
  - Urban green areas

- Supporting services

  - Nationally protected areas/habitats, protected species
  - International status: Ramsar convention, UNESCO Man and Biosphere, World Heritage Sites
  - Subject to national (e.g., UK Biodiversity Action Plans) or regional regulations (e.g., European Union Natura 2000 Network)
  - Sites hosting species listed under the Convention on the Conservation of Migratory Species of Wild Animals of the Convention on International Trade in Endangered Species of wild Flora and Fauna
  - Sites hosting species listed under the Bern Convention

Secondly, identify the specific ES-related content of the identified regulations, and present it in a way that can be easily communicated to policy makers and stakeholders. This may include producing maps showing areas of concern for the specific regulation (e.g., designated sites, buffer zones of water bodies, habitat maps) or summaries of key elements (e.g., minimum requirements for green space in urban areas, no-net-loss policy on pervious surfaces, constraints on land development). Thirdly, provide initial comments (as far as it is feasible at this stage) on the implications of the regulation for the development of the plan. The latter action involves answering questions such as:

- What geographical areas/ecosystem types are addressed by the regulation?
- What stakeholders and beneficiary groups, disaggregated by gender and other sensitive groups, are mainly concerned?
- Does the regulation set constraints to decision making? How?
- Does the regulation offer opportunities for synergy with the plan? How can the plan contribute to the regulations objectives and vice versa?
- What specific elements of the plan are concerned the most by the regulation?

In conclusion, the output of this task should not be a mere listing of existing regulations: this would simply add a layer to the huge pile of information that decision makers should be aware of, with likely limited effects on the final outcome. A further step needs to be taken, by identifying key content and bringing it to the attention of decision makers in a clear and concise way, together with comments on the potential synergies and criticalities. In this way, the output can serve the purpose of both reminding decision makers of issues that need to be taken into account (in a "reactive" way) and proposing ideas and strategies (in a "proactive" way). Obviously, synergies and constraints can be more or less identifiable according to the state of advancement of the plan. For this reason, the output of this task is not intended as a static picture, but needs to be updated and revised during the SEA, and used to inform the process.

Task 1.3: identify links with other strategic actions

This task aims at identifying other relevant strategic actions at various levels (e.g. national, regional, local) whose content must be taken into account to exploit synergies and reduce inconsistencies in terms of ES use and conservation. The task is similar to the previous one in that its purpose is to harmonize the plan with the external context. Even though the analysis is typically carried out for external actions that belong to higher or equal decision levels (e.g. for a regional strategic action: national and regional Policies, Plans and Programmes [PPP]), it can be worth exploring also actions on lower levels (e.g., local-level PPP), as well as individual projects. In particular, large-scale projects (e.g., a dam, a major transportation infrastructure) may influence the content and implementation of the plan.

In SEA, this task is often called "external compatibility appraisal". It can be conducted both in a reactive (i.e., by testing if the proposed plan is compatible with the external context) and proactive way (i.e., by using information on the external context to shape the content of the plan and exploit synergies). Guiding questions to identify links with other strategic actions include:

- Do the objectives of other plans or policies depend on ES that will be affected by the plan?
- Are other plans or policies likely to affect ES that are needed to achieve the objectives of the plan?
- Does the plan contribute to enhance ES that are needed by an external plans or policies, or vice versa?

This task needs to be repeated throughout the planning process. In the preliminary stages it is conducted by looking at the objectives proposed in the plan. Later on it can be performed by analyzing the specific policies and activities proposed to achieve such objectives. This reiteration is important because objectives might be too broad or too vague to allow a proper understanding of their effects on ES. Specific policies (e.g., a zoning scheme for a spatial plan) will allow unveiling of critical interactions that can be brought to the attention of the decision makers at a stage where they can still be corrected or reviewed.

Stage 2: determine and assess priority ES

The output of Stage 1 is likely to include an extensive list of ES and associated beneficiary groups. In order to improve the effectiveness of SEA, Stage 2 generates detailed information on a limited set of ES, which are considered relevant for shaping and informing the development of the plan. This requires determining priority ES (Task 2.1), and assessing their baseline conditions and trends (Task 2.2).

Task 2.1: determine priority ES

Setting priority ES is best done in close collaboration with stakeholders and beneficiaries. In these initial stages, the content of the plan is typically in the form of a draft set of problems that the plan wishes to solve, and objectives that it wishes to achieve. By analysing this content, a preliminary screening can be performed in order to identify:

- The services upon which the plan depends
- The services that the plan may affect (positively or negatively)

A plan depends on an ES if the service is an input or if it enables, enhances or regulates the conditions necessary for a successful outcome of the plan (OECD, 2008). For instance, a tourism development plan may depend upon cultural services (such as aesthetic value) provided by coastal ecosystems. A plan affects an ES if it triggers drivers that decrease (negative impact) or enhance (positive impact) the quantity or quality of that service. For instance, a regional development plan may promote land-use changes that negatively affect the provision of freshwater. This systematic analysis of dependences and impacts helps uncover unforeseen interactions between ES and the plan. Identifying these interactions up front will enable decision makers to proactively manage any associated risks and opportunities (Ranganathan et al., 2008).

Guiding questions to support the identification of priority ES include (OECD, 2008; Landsberg et al., 2013):

- Is the plan likely to trigger or reinforce drivers that contribute to the degradation of the ecosystems?
- Can the plan trigger or reinforce drivers that contribute to enhancement of ES important for people's well-being (e.g. by improving the quality and quantity of ES supply or by enhancing the ability of people to benefit from ES)?
- Is the plan likely to limit the ability of people (within and outside the planning region) to benefit from ES?
- Is the plan likely to affect the demand for a given ES, either directly (because the plan depends on it for the achievement of its objectives) or indirectly (because it increases demand by other)?
- Is economic development and human well-being for different groups of people likely to be affected by a decline in the ES?
- Is the affected ES a major contributor to the well-being of the potentially affected groups of people?
- Does the affected ES have a cost-effective substitute?

Table 2.2 shows an example of analysis of dependencies and impacts to identify priority ES. Specifically, it presents the analysis of the possible relationships

Table 2.2 Analysis of dependencies and impacts to identify priority ES for the SEA of a spatial plan

	To protect watersheds		To ensure suitable development of quarrying, housing, transportation and agriculture activities								To develop a non-invasive eco-tourism industry	
			Quarrying		Housing		Transport		Agriculture			
	D	I	D	I	D	I	D	I	D	I	D	I
Soil conservation	●	++		--		-		-	○	+/-		-
Water regulation	●	++		-	●	--			●	--	●	
Hunting		+		-				-		-		
Water purification		++		--	●	-				--		
Carbon sequestration		++		-		--						
Recreation		+/-		-				+/-		+	○	++
Aesthetic appreciation		+/-		-	●					+	○	+/-
Food and medicine		-/+		-		-		+		++		

Key: D: Dependence; I: Impact; (+)+: (Very) Positive impact; (-)-: (Very) Negative impact; ●: Dependence between objective and ES; ○: Dependence that extends beyond the planning region.

Source: Geneletti (2016b)

Note: The matrix has been adapted from the one produced by the participants of the workshop "Integrating ecosystem services in Strategic Environmental Assessment for policy support", held in Port of Spain on December 3–5, 2014. All elements of this matrix are purely hypothetical and intended for illustrative purposes only.

between some of the objectives of a hypothetical Spatial Plan for the development of a mountain region in Trinidad and Tobago and ES. For each of the three plan objectives, the first column indicates the ES required for its achievement. The matrix indicates also when such dependence may extend beyond the boundary of the area being planned, hence requiring a broader scale analysis (e.g., the regulation of water may depend upon decisions taken outside the region). The second column identifies situations where the achievement of the objective will have a positive/negative effect on the ES. For instance, the protection of natural areas is bound to contribute to soil formation and retention, but it may have both positive and negative effects on recreation opportunities. Similar analyses are useful also to test the "internal consistency" of the plan. Potential inconsistencies exist whenever the achievement of one objective relies on a given service, which can be affected by a different objective. These situations can be detected by looking at each row of the matrix (see, for instance, the case of water regulation). The results of this type of analyses can suggest, for example, revisions of the objectives, but also additional stakeholders to be consulted (i.e., beneficiaries of the services affected).

This task requires wide stakeholder consultation to obtain existing information and to confirm the values, interests and dependencies on priority ES with people who need and use them, considering also gender issues. Stakeholder engagement is crucial to understand the complex relationships between a society and its biophysical environment. An effective stakeholder consultation ensures that no relevant issues are left out, and allows decision makers to properly set the "boundaries" of the SEA in a way to encompass the views and interests of all affected people.

A critical issue in stakeholder consultation is represented by the involvement of the poor. Biodiversity has been described as "the wealth of the poor" (WRI, 2005), but power imbalances and governance failures make the poor often invisible and not fully involved in the planning/policy-making processes concerning the use of natural resources. This, together with problems such as weak land rights, weakly enforced legislation and corruption cause the benefits of ecosystems to be captured by those far away (e.g., genetic resources exploited by international corporations) or by national governments with limited local effects (e.g., wildlife tourism) to the detriment of the poor, who are stewards of the ecosystems (Roe et al., 2011). This issue needs to be seriously considered in SEA, by improving participation of the less wealthy and more vulnerable groups (e.g., minorities) in the identification of priority ES (as well as in subsequent decision-making stages), in order to ensure that their interests are not overridden in favor of more powerful concerns.

Task 2.2: assess baseline conditions and trends for priority ES

In this task, a detailed analysis of the current state of priority ES, as well as their likely evolution without the plan, is carried out. The output should provide as clear a picture as possible about:

- Current distribution of priority ES, and benefits provided to different groups of people

- Key direct and indirect driving forces
- Likely future trends, threats and opportunities

Assessing baseline conditions for ES can be challenging or overly time consuming. However, one must remember that baseline data in SEA essentially serve three purposes (adapted from Therivel, 2004): (1) identifying critical issues and opportunities related to ES to ensure that they can be addressed by the plan; (2) describing current conditions and expected trends so as to have a reference against which measuring the performance of the plan and (3) providing a basis for the prediction and assessment of the impact on ES. These purposes should be kept in mind when deciding when to stop collecting and processing baseline data, and move on. A complete baseline is not necessarily needed to proceed with the SEA, and additional data should be collected only if they provide a relevant contribution to one of the previous purposes.

The assessment of ES can be conducted in qualitative or quantitative way. Quantitative assessments, in turn, can be based on monetary or non-monetary (e.g., biophysical) measures. The complementarity of different assessment approaches should be acknowledged in SEA practice. Monetary valuation offers many advantages, but it may not be always appropriate or even possible (TEEB, 2013). The assessment of ES in their own terms may be more meaningful for stakeholders than a monetary value (e.g., the recreational or spiritual value of a landscape feature). Hence, different types of assessment can be chosen for the different ES. The objective and scope of the SEA (including the foreseen interactions with stakeholders along the process), as well as the availability of data, time and resources, will play a key role in selecting the appropriate way to assess ES, as well as the specific methods.

It is fundamental for this task to include a dynamic component, by providing information not only about the current conditions, but also (and especially) about possible future trends. This will provide the basis for developing the plan in a way that it can adequately "fit" these trends (e.g., by reducing risks and exploiting opportunities related to ES). It will also provide the basis for assessing the impact of the plan against the baseline conditions (see Stage 3). Analysing trends in ES requires the identification of key drivers that are influencing them. Drivers can be of a direct nature (e.g., physical interventions, such as land use changes) or an indirect one (policies that may affect the way in which society makes use of ES, such as for instance the ones that regulate accessibility to recreation areas). Operational guidance on the analyses of drivers and trends can be found in UNEP (2009) and Ash et al. (2010).

Stage 3: identify alternatives and assess impacts on ES

In this stage, the plan is taking shape and specific alternatives are proposed. SEA has the purpose of contributing to the identification of possible alternatives to enhance ES, or at least minimizing negative effects on them (Task 3.1), predicting and evaluating impacts for each alternative (Task 3.2), and identifying measures to enhance and mitigate impacts (Task 3.3).

Task 3.1: identify alternatives

This task has the purpose of contributing to the identification of possible courses of action to enhance (priority) ES, or at least minimize negative effects on them. The analyses conducted in the previous two stages and the information collected so far (including stakeholders' perceptions and values) is used to ensure that key ES-related issues are mainstreamed in the actual content of the plan. Alternatives can be generated as a reaction to proposals formulated by planners/policy makers (e.g., proposing infill development as opposite to urban expansion in areas that provide important water regulation services), or as a response to issues that emerged during the previous stages, and that need to be adequately addressed by the plan (e.g., proposing a constraint to land-use conversion in an area that proved to be essential in providing a priority ES). Alternatives developed during the SEA can be radically different ways to achieve a given objective, or can result from adjusting and fine-tuning existing proposals (e.g., by suggesting better implementation details or location for a given activity).

Ideally, the proposed alternatives should contribute to:

*   Promoting synergy (or at least avoiding conflict) with existing regulations, plans and policies relevant for ES
*   Minimizing the dependency of the plan on priority ESs
*   Enhancing positive (and minimizing negative) effects on priority ES
*   Having desirable effects on drivers of ES change
*   Generating a more equitable distribution of cost and benefits, particularly considering vulnerable stakeholders

Task 3.2: predict and evaluate impacts

This task has the purpose of providing information on which ES would benefit or be worse off, and which groups of people would win or lose, if a given alternative is selected. This information will provide the basis for discussions with stakeholders and planning/policy makers and for supporting the final decision-making process (which typically requires knowledge also on additional issues besides ES). The identified alternatives are compared in terms of their impact on ecosystem services, in order to suggest the options that enhance opportunities (e.g., for conservation of ES, improvement of quality/quantity of ecosystem service provision, increase in potential beneficiaries) and reduce risks (related, for example, to high level of dependence on ESs, degradation, conflict in access and use).

Impacts are defined as the difference between the conditions of a given variable with and without the plan through time. Hence, impacts can be desirable (positive) or undesirable (negative) changes that result from the implementation of the plan. Impact assessment should inform about the consequences of the alternatives under consideration on the provision of a given service (e.g., change in quality/quantity of yield of crop; change in denitrification capability within a watershed; change in the area of landscape in attractive condition). It should also inform about the

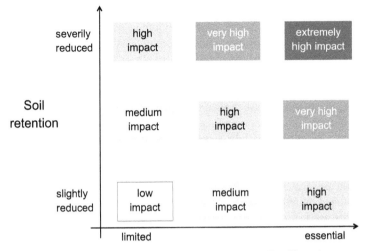

Figure 2.4 Conceptual diagram to assess the significance of ES impact. The impact significance is estimated by combining the expected magnitude of change in the capability of ecosystems to prevent erosion (y-axis: soil retention), and the importance of well-preserved soil to the livelihoods of local villages (x-axis: contribution of farming to livelihoods). The highest impacts correspond to severe reductions in soil retention in areas where people rely on subsistence farming.

Source: Geneletti (2016b)

importance of such changes for beneficiaries, by considering issues such as dependency and replaceability, poverty, vulnerability, access, and so on (see Landsberg et al., 2013). Hence, the overall significance of an impact is a function of both the magnitude of the change, and the importance of that change for the well-being of the affected people (Figure 2.4).

Performing impact assessment requires knowledge of the relationship between ecosystems, ES and human well-being, which was gained in the previous stages. In particular, the baseline and trends analysis provides the reference against which the performance of different alternatives can be measured (including the "do-nothing" alternative, if meaningful), and the basis for suggesting the alternative(s) that better fit the desirable future conditions that the plan is pursuing (Geneletti, 2013b).

Several methods and techniques for impact assessment can be used, according to the level of detail of the analysis and the way in which ES have been characterised in the baseline (e.g. models and quantitative analysis, expert opinion and qualitative descriptions, monetary evaluation, assessment of ES in their own terms, etc. See examples in Burkhard et al., 2013). Generally, in SEA qualitative impact, prediction and evaluation are more common than quantitative ones, due to the inherently high uncertainty levels (in the data, in the future trends of key drivers, etc.), the complexity of the decisions and the time and resources constraints of the planning/policy-making exercise (Jones et al., 2005). However, examples of more quantitative approaches for ecological impact assessment in SEA exist (e.g. Noble, 2008). Even though issues addressed by SEA are in general less tangible than those addressed at the project level, there is a lot of room to improve the analytical content of SEA, as argued by Geneletti (2015b). This applies particularly to plans and programmes that provide detailed regulations, such as for example the zoning scheme of an urban plan that identifies permitted/prohibited land-use changes in each land unit.

Whenever possible, spatially explicit impact assessment methods should be preferred because they offer a better understanding of the complex relationships between areas of ES production and use, and they may help to differentiate impacts by beneficiary groups (e.g., mountain versus lowland villages; wealthy versus disadvantaged neighbourhoods; different municipalities within a region). Spatial approaches are in general computationally more complex and need more data. However, most baseline data related to ES are typically available in map format, and new software tools are being developed that use relatively simple models with few input requirements (Kareiva et al., 2011). Qualitative spatial approaches (e.g., participatory mapping) can be particularly useful to engage stakeholders and communicate results.

Many problems related to the loss or degradation of ES result from the cumulative effects of human activities. Cumulative effects are the net impact from a number of different activities and can occur from the following situations (Cooper, 2004):

- Interaction of impacts from proposals and policies within a plan affecting the same ES. For example, proposals to build infrastructures, commercial premises and housing within a short period of time could result in cumulative loss of open space and attractive landscape for recreation. Analogously, a policy to encourage renewable biofuels cultivation and a land consolidation policy could result in a cumulative loss of subsistence cropping.
- Combined impacts of the strategic action with impacts of other actions affecting the same ES in a particular area. For example, proposals from urban and forest plans could interact and affect the regulation of local climate.

One of the main raison d'être of SEA is the assessment of cumulative effects, given that individual impacts from a single project or development may not be significant on their own, but become significant in combination with other impacts. Hence, SEA cannot be limited to the analysis of individual elements of the plan, but needs to also carry out an overall assessment of the future conditions of priority ES, in the light of all the activities and policies that the plan includes. Additionally, other past, present and reasonably foreseeable future actions within space and time boundaries that could contribute to cumulative effects on a given ES should be considered (Canter and Ross, 2010). Cumulative effects can also be positive. For example, economic incentives for planting hedgerows and trees in rural areas

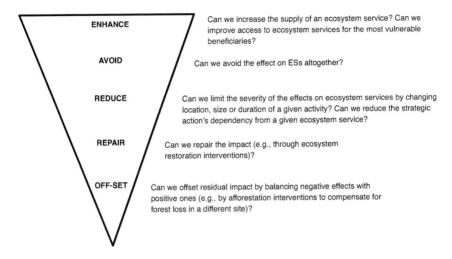

Figure 2.5 The "mitigation hierarchy" and examples of guiding questions to identify measures for enhancement and mitigation of the plan effects on ES.

and policies to promote riverbanks restoration could cumulatively result in better nutrient retention and water filtration.

Task 3.3: identify measures to enhance and mitigate impacts

This task aims at suggesting how to enhance positive impacts and opportunities connected to the implementation of the plan and mitigate negative impacts and risks. Enhancement and mitigation measures may include changes to the plan (e.g., removal/addition/refinement of elements, such as policies or regulations), as well as guidelines for later decisions. The latter comprise, for example, recommendations for institutional adaptation or new regulations that should be taken onboard in subsequent policies or plans and recommendations for project's EIA (e.g., terms of reference for future EIA of projects affecting a specific area or ES) (Partidário, 2012).

Following the revised mitigation hierarchy proposed by Bond et al. (2013), the SEA should seek measures that, in order of priority: enhance ES; avoid negative effects on ES; reduce negative effects; repair negative effects; offset negative effects. Figure 2.5 illustrates this concept and provides some useful guiding questions to identify measures to enhance and mitigate impacts.

Stage 4: follow up on ES

This stage begins when the plan has been approved and aims at understanding the effective progress in its implementation, the actual impacts on ES, as well as relevant contextual changes. It entails two tasks: monitoring and managing ES during implementation and testing the quality of the SEA process.

Task 4.1: monitor changes in the context and impacts on ES

Follow-up is defined as the monitoring and evaluation of the impacts of the plan for management of, and communication about, its environmental performance (Morrison-Saunders and Arts, 2004). Follow-up ensures continuity to the SEA process, and allows detecting contextual changes that may occur during the implementation of the plan, along with progress in its effective implementation and actual impacts (Partidário, 2012).

This task has two main objectives: (1) monitoring changes in the ES context; (2) monitoring the actual impacts of the plan on ES. Concerning the first objective, insights need to be gained into the environmental and socio-economic changes that occur in the area during the implementation of the plan, and that may affect the supply, demand or use of ES. Relevant changes that have direct or indirect effects on ES can be related to the state of the environment (e.g., climate trends, natural disasters), the social profile (e.g., migration patterns), the socio-economic situation (e.g., shift in livelihood systems), and the legislative and regulatory framework (e.g., designation of protected new areas; land reform policy). Existing monitoring and reporting programmes (e.g., state of the environment reports) represent a useful starting point for this activity. Early detection of relevant contextual changes enables adjustments to the plan, thus ensuring the continuing facilitating role of SEA (Partidário, 2012). This task is also instrumental to filling the gap in knowledge (e.g., on the quantification of ES) that arose during the SEA, but could not be addressed for lack of time or resources.

The following are examples of guiding questions that can be used to periodically monitor the context and detect changes that need to be brought to the attention of the people in charge of implementing or revising the plan:

- Have new formal regulations linked to ES been approved in the study region (e.g., designated areas, PES schemes)?
- Have other policies, plans or projects been approved or implemented that could affect the supply, demand or use of ES (e.g., energy policy, agricultural reform, urban plan)?
- Has the demand for a particular priority ES by stakeholders changed (e.g., due to droughts, change in trade policy, change in access to specific locations)?
- Has the supply of a particular priority ES changed (e.g., following a natural disaster)?
- Is there new evidence available concerning ES (e.g., ecosystem maps, economic valuations, surveys on users' needs)?
- Does this new evidence or knowledge suggest changes in the analysis carried out during the SEA (e.g., assessment of baseline conditions and future trends, impact prediction), hence in the final recommendations?

With respect to the second objective, evidence needs to be collected about the actual impacts of the plan on ES, in order to evaluate to what extent the observed impacts differ from the predictions performed during the previous stages. The ultimate purpose is to enable timely intervention and correction of detected problems (e.g., unforeseen impacts, mitigation and enhancement measures not carried out). Important elements of an efficient monitoring system include (Morrison-Saunders and Arts, 2004):

- Identification of indicators and description of the methods, frequency and responsibility for data collection (responsibility may vary according to the nature of the indicators)
- Guidance on data evaluation (appraisal of the conformity with predictions or expectations)
- Guidance on management (how to take appropriate response to issues arising from evaluation)
- Communication strategies (informing stakeholders about the results or, where appropriate, involving stakeholders in the evaluation and/or management stages)

The selected indicators should be consistent with the ones used to inform the previous SEA analyses (e.g., indicators used in Task 2.2), limited in number (to ensure viability of the monitoring system), and possibly easy to measure, interpret and communicate. Besides contributing to a better implementation of the plan under consideration, this task has an important learning objective for future SEA: it helps understanding what went wrong with impact prediction and how to make better predictions, hence improving the practice of mainstreaming ES into SEA and decision-making.

Task 4.2: test the quality of the SEA

This task aims at checking if the SEA process has been carried out well, with respect to consideration of ES. Being the process that matters, and not so much the content of the final report, it is recommended that policy makers perform this analysis throughout the SEA, rather than only at the end of it. In this way, short-comings and limitations can be highlighted, and integration proposed when they can materially be used to improve the plan. The lessons learned from quality control checks are also beneficial for future applications and can improve the practice of integrating ES into SEA. This is currently a very important issue, given the lack of experiences and case studies in this field (Rosa and Sánchez, 2015).

The following guiding questions can be used to check the influence of ES information on the planning process and content:

- Was the information on ES provided by the SEA process adequate and useful from the point of view of both decision makers and stakeholders?
- Has there been effective cooperation on ES issues between the SEA team and those responsible for developing the plan?
- Was there effective stakeholder involvement on issues relevant to ES?
- Did the SEA lead to measures and outcomes that better reflect ES in the planning/policy making process?

- Did the SEA succeed in integrating into the plan operational measures (e.g., budget allocation) for dealing with risks of depleting ES?
- What were the main strengths and weaknesses of the SEA process (in terms of availability of data on ES, analysis of ES relevance, stakeholder involvement, etc.)?
- Did the SEA improve the capacities of decision makers and stakeholders to manage ES?
- Did the SEA enhance the transparency and accountability of the planning decisions related to ES?
- Did the plan contribute to verifiable progress on ES protection/enhancement?

Finally, the following guiding questions can be used to check the extent to which the SEA content reflected knowledge of ES:

- Did the SEA identify priority issues for ES, rather than all potentially significant issues?
- Have the substantial objectives related to ES conservation/enhancement been identified and described?
- Did the SEA identify and describe any conflicts that exist between these objectives and the plan?
- Did the SEA take into account alternative options, based on the way these alternatives affect ES?
- Did the SEA provide useful information on ES-related risks/opportunities related to the plan, and on mitigation measures/adaptive strategies that could be adopted?
- Were the impacts, and the methodologies for assessing impacts, on ES clearly described?

Conclusion

This chapter provided guidance to include in SEA analyses that can clarify the potential impacts of planning decisions on ES, in order to avoid unintended negative consequences and seize opportunities for improvement. The proposed analyses can be used in different contexts and for different types of SEA processes. The integration of ES in SEA has various benefits in terms of contributing to better design of policies and plans, but there are also critical issues that need to be recognized. These include the complexity of appropriately evaluating ES, and the lack of well-established indicators and assessing methods. These issues can be addressed by learning from the pilot applications and case studies that are being carried out around the world, as well as by taking stock of the data, tools, and methods for ES analysis that are becoming increasingly available in the scientific (and grey) literature (Guerry et al., 2015). Practitioners are faced with the challenge of including ES and showing their added value to decision making, within the time and resource constraints of real-life planning and policy making processes.

Acknowledgements

This chapter draws on the guidance manual "Integrating Ecosystem Services in Strategic Environmental Assessment: A Guide for Practitioners", produced by the author for the UNEP's GEF-funded Project for Ecosystem Services, implemented by the Ecosystem Services Economics Unit, DEPI-UNEP. Different versions of this chapter, based on the same guidance manual, appeared in the *Journal of Environmental Assessment, Policy and Management* (Geneletti, 2015a) and in the *Handbook of Biodiversity and Ecosystem Services in Impact Assessment*, Edward Elgar Publishing (Geneletti, 2016b).

References

Abaza, H., Bisset, R., and Sadler, B. (2004). *Environmental impact assessment and strategic environmental assessment: Towards an integrated approach.* UNEP, Geneva.

Ahmed, K. and Sánchez-Triana, E. (eds.) (2008). *Strategic environmental assessment for policies.* The World Bank, Washington, DC.

Ash, N., Blanco, H., Brown, C., Garcia, K., Henrichs, T., Lucas, N., . . . Zurek, M. (2010). *Ecosystems and human well-being: A manual for assessment practitioners.* Island Press, Washington, Covelo, and London.

Baker, J., Sheate, W. R., Phillips, P., and Eales, R. (2013). Ecosystem services in environmental assessment: Help or hindrance? *Environmental Impact Assessment Review*, 40, 3–13.

Bond, A., Morrison-Saunders, A., and Stoeglehner, G. (2013). Designing an effective sustainability assessment process. In: A. Bond, A. Morrison-Saunders and R. Howitt (eds.) *Sustainability assessment pluralism, practice and progress.* Routledge, London and New York, 231–244.

Burkhard, B., et al. (eds.) (2013). Mapping and modelling ecosystem services for science, policy and practice: Special issue. *Ecosystem Services*, 4, 1–146.

Canter, L. and Ross, B. (2010). State of practice of cumulative effects assessment and management: The good, the bad and the ugly. *Impact Assessment and Project Appraisal*, 28(4), 261–268.

Cooper, L. M. (2004). Guidelines for Cumulative Effects Assessment in SEA of Plans. EPMG Occasional Paper 04/LMC/CEA, Imperial College London.

Fischer, T. B. (2007). *Theory and practice of strategic environmental assessment: Towards a more systematic approach.* Earthscan, London.

Geneletti, D. (2011). Reasons and options for integrating ecosystem services in strategic environmental assessment of spatial planning. *International Journal of Biodiversity Science, Ecosystem Services & Management*, 7(3), 143–149.

Geneletti, D. (2013a). Ecosystem services in environmental impact assessment and strategic environmental assessment: Special issue. *Environmental Impact Assessment Review*, 40, 1–87.

Geneletti, D. (2013b). Assessing the impact of alternative land-use zoning policies on future ecosystem services. *Environmental Impact Assessment Review*, 40, 25–35.

Geneletti, D. (2015a). A conceptual approach to promote the integration of ecosystem services in strategic environmental assessment. *Journal of Environmental Assessment Policy and Management*, 17(4).

Geneletti, D. (2015b). Research in strategic environmental assessment needs to better address analytical methods. *Journal of Environmental Assessment Policy and Management*, 17(1).

Geneletti, D. (2016a). *Handbook on biodiversity and ecosystem services in impact assessment*. Edward Elgar, Cheltenham, UK and Northampton, MA.

Geneletti, D. (2016b). Ecosystem services analysis for strategic environmental assessment: Concepts and examples. In: D. Geneletti (ed.) *Handbook on biodiversity and ecosystem services in impact assessment*. Edward Elgar, Cheltenham, UK and Northampton, MA.

Geneletti, D., Bond, A., Russel, D., Turnpenny, J., Sheate, W., and Jordan, A. (2015). Ecosystem services and sustainability assessment: Theory and practice. In: A. Morrison-Saunders, J. Pope and A. Bond (eds.) *Handbook of sustainability assessment*. Edgar Elgar, Cheltenham, UK, 215–134.

Guerry, A. D., Polasky, S., Lubchenco, J., Chaplin-Kramer, R., Daily, G. C., Griffin, R., . . . Vira, B. (2015). Natural capital and ecosystem services informing decisions: From promise to practice. *Proceedings of the National Academy of Sciences*, 112(24), 7348–7355.

Haines-Young, R. H. and Potschin, M. P. (2010). The links between biodiversity, ecosystem services and human wellbeing. In: D. G. Raffaelli and C. L. J. Frid (eds.) *Ecosystem ecology: A new synthesis*. Cambridge University Press, Cambridge, 110–139.

Helming, K., Diehl, K., Geneletti, D., and Wiggering, H. (2013). Mainstreaming ecosystem services in European policy impact assessment. *Environmental Impact Assessment Review*, 40, 82–87.

Honrado, J. P., Vieira, C., Soares, C., Monteiro, M. B., Marcos, B., Pereira, H. M., and Partidário, M. R. (2013). Can we infer about ecosystem services from EIA and SEA practice? A framework for analysis and examples from Portugal. *Environmental Impact Assessment Review*, 40, 14–24.

IAIA (2002). Strategic environmental assessment: Performance criteria. International Association for Impact Assessment Special Publication Series No. 1. Available from: www.iaia.org/publicdocuments/special-publications/sp1.pdf.

IPBES (2013). Recommended conceptual framework of the intergovernmental science-policy platform on biodiversity and ecosystem services. Working Document IPBES/2/5. Available from: http://ipbes.net/images/K1353197-en.pdf.

Jones, C., Baker, M., Carter, J., Jay, S. J., Short, M. B., & Wood, C.J. (2005). *Strategic environmental assessment and land use planning : an international evaluation*. Earthscan London.

Kareiva, P., Tallis, H., Ricketts, T. H., Daily, G. C., and Polasky, S. (2011). *Natural capital: Theory and practice of mapping ecosystem services*. Oxford University Press, Oxford.

Karjalainen, T. P., Marttunen, M., Sarkki, S., and Rytkönen, A.-M. (2013). Integrating ecosystem services into environmental impact assessment: An analytic-deliberative approach. *Environmental Impact Assessment Review*, 40, 54–64.

Kumar, P., Esen, S. E., and Yashiro, M. (2013). Linking ecosystem services to strategic environmental assessment in development policies. *Environmental Impact Assessment Review*, 40, 75–81.

Landsberg, F., Treweek, J., Stickler, N. M., and Venn, O. (2013). *Weaving ecosystem services into impact assessment*. World Resource Institute, Washington, DC.

MA (Millennium Ecosystem Assessment) (2005). *Ecosystems and human well-being: The assessment series (four volumes and summary)*. Island Press, Washington, DC.

Maes, J., Teller, A., Erhard, M., Liquete, C., Braat, L., et al. (2013). *Mapping and assessment of ecosystems and their services: An analytical framework for ecosystem assessments under action 5 of the EU biodiversity strategy to 2020*. Luxembourg, Publications Office of the European Union.

Mandle, L., Bryant, B. P., Ruckelshaus, M., Geneletti, D., Kiesecker, J. M., and Pfaff, A. (2015). *Entry points for considering ecosystem services within infrastructure planning: How to integrate conservation with development in order to aid them both.* Conservation Letters.

Morrison-Saunders, A. and Arts, J. (2004). Introduction to EIA Follow-up. In: A. Morrison-Saunders and J. Arts (eds.) *Assessing impact: Handbook of EIA and SEA follow-up.* Earthscan, London, 1–21.

Noble, B. (2008). Strategic approaches to regional cumulative effects assessment: A case study of the Great Sand Hills, Canada. *Impact Assessment and Project Appraisal*, 26(2), 78–90.

OECD (2006). *Applying strategic environmental assessment: Good practice guidance for development co-operation.* DAC Guidelines and Reference Series. Organisation for Economic Cooperation and Development, Paris.

OECD (2008). *Strategic environmental assessment and ecosystem services: DAC Network on Environment and Development Cooperation (ENVIRONET).* Organisation for Economic Cooperation and Development, Paris.

Partidário, M. R. (2012). *Strategic environmental assessment better practice guide: Methodological guidance for strategic thinking in SEA.* Portuguese Environment Agency and Redes Energéticas Nacionais, Lisbon.

Partidário, M. R. and Gomes, R. C. (2013). Ecosystem services inclusive strategic environmental assessment. *Environmental Impact Assessment Review*, 40, 36–46.

Ranganathan, J., Raudsepp-Hearne, C., Lucas, N., Irwin, F., Zurek, M., Bennett, K., . . . West, P. (2008). *Ecosystem services: A guide for decision makers.* World Resource Institute, Washington, DC.

Roe, D., Thomas, D., Smith, J., Walpole, M., and Elliott, J. (2011). Biodiversity and poverty: Ten frequently asked questions: Ten policy implications: International institute for environment and development. *Gatekeeper*, 150.

Rosa, J. C. S. and Sánchez, L. E. (2015). Is the ecosystem service concept improving impact assessment? Evidence from recent international practice. *Environmental Impact Assessment Review*, 50, 134–142.

Sadler, B. (2011). Taking stock of SEA. In: B. Sadler, R. Aschemann, J. Dusik, T. B. Fischer, M. Partidario, and R. Verheem (eds.) *Handbook of strategic environmental assessment.* Earthscan, London, 1–19.

Slootweg, R., Kolhoff, A., Verheem, R., and Höft, R. (2006). *Biodiversity in EIA and SEA: Background document to CBD decision VIII/28: Voluntary guidelines on biodiversity-inclusive impact assessment.* Commission for Environmental Assessment, The Netherlands.

TEEB (2011). *The economics of ecosystems and biodiversity in national and international policy making.* Edited by P. ten Brink. Earthscan, London and Washington, DC.

TEEB (2013). Guidance manual for TEEB country studies: The economics of ecosystems and biodiversity. Version 1.0. Available from: www.unep.org/pdf/TEEB_GuidanceManual_2013.pdf (last access: 08/09/2015).

Therivel, R. (2004). *Strategic environmental assessment in action.* Earthscan, London.

UNEP (2009). *Integrated assessment for mainstreaming sustainability into policymaking: A guidance manual.* UNEP, Geneva.

World Bank (2009). *Strategic environmental assessment in policy and sector reform: Conceptual models and operational guidance.* The World Bank, Washington, DC.

WRI (2005). *The wealth of the poor: Managing ecosystems to fight poverty.* World Resources Institute, Washington, DC.

3 Lessons for mainstreaming ecosystem services into policy and practice from South Africa

Nadia Sitas, Jeanne Nel and Belinda Reyers

Introduction

The Southern African Millennium Ecosystem Assessment (SAfMA), which took place between 2001 and 2004, catalysed many subsequent ecosystem service research projects, assessments and policy development in the region (Biggs et al. 2004; Bohensky et al. 2004; Bohensky & Lynam 2005; Shackleton et al. 2004; Van Jaarsveld et al. 2005). Several of these successive efforts were specifically targeted at further integrating (mainstreaming) ecosystem services and ecosystem assessments into multiple spheres and scales of policy and practice in South Africa. For the purposes of this chapter, mainstreaming refers to the informed integration of *"relevant environmental concerns into the decisions of institutions that drive national, local and sectoral development policy, rules, plans, investment and action"* (Dalal-Clayton & Bass 2009, emphasis in original). Mainstreaming efforts, strengthened by the initiation of the Project for Ecosystem Services (ProEcoServ) in southern Africa, sought to build partnerships with various stakeholders including government, private sector, and civil society, with an aim of having a lasting impact on ecosystem management and its role in multi-scale development planning and decision making.

This collaborative work relied on various strategies for mainstreaming ecosystem services, involving aspects such as communication and outreach, co-production of knowledge, integration of ecosystem services into national policy and dialogue (such as through co-producing national ecosystem-service maps), promotion of public-private cooperation for ecosystem management, development of ecosystem-service models to inform investments, and framework development for investment in ecological infrastructure that in turn has been adopted in national policy and planning.

Approach

Throughout this work there was an effort to collate mainstreaming lessons from an African perspective while also interacting with international stakeholders and agencies to test and disseminate these to global science and policy platforms. This was accelerated by the initiation of the Sub-Global Assessment network (www.ecosystemassessments.net/) following the Millennium Ecosystem Assessment

(MA 2005), and the later established Intergovernmental Science-Policy Platform on Biodiversity and Ecosystem Services (IPBES, see Díaz et al. 2015 and www. ipbes.net). Both these processes have a strong regional focus on Africa, providing opportunities for distilling lessons on mainstreaming but also as a training tool for thinking more explicitly about what makes for effective mainstreaming and actionable science.

Our contribution in these processes focused on: (1) developing and documenting a mainstreaming framework based on ProEcoServ case studies, emerging concepts from the literature and common lessons explored during our outreach; (2) refining the mainstreaming framework through workshops and (3) communicating the mainstreaming framework to global science and policy networks.

ProEcoServ case studies

CASE STUDY 1: DISASTER MANAGEMENT IN THE EDEN DISTRICT – PROMOTING DISASTER RESILIENCE THROUGH ECOSYSTEM-BASED MANAGEMENT

Introduction

Disasters caused by natural hazards – like floods, droughts, wildfires and storm-waves – have been responsible for the loss of at least a million lives over the last decade, with recovery often taking years and financial losses estimated to be in the trillions of US dollars.[1] What's more – the frequency and intensity of these natural hazards is rapidly increasing. The causes for these increases are often attributed to climate changes as well as changes in the patterns of human exposure. However, there is also a growing concern that rapid and widespread land-cover change is leading to the loss of the buffering capacity that healthy ecosystems and their services provide against natural hazards.

The Eden District – a mountainous, biodiversity rich area in the south coastal region of South Africa – is particularly vulnerable to such hazards. The region receives rainfall throughout the year, with peaks in March and October, often associated with cut-off low events in southern Africa. The area's rainfall pattern and mountainous nature make it prone to flash floods followed by droughts. Large storm-waves often occur with high rainfall events resulting in severe coastal flooding. The region is located within the fynbos biome, a fire-prone vegetation type, which makes it further vulnerable to large wildfires that are exacerbated by dense infestations of non-native invasive shrubs and trees that increase the frequency and intensity of wildfires.

Future increases in extreme events are predicted in the Eden District linked to predicted climate changes including: a 2°C increase in temperature by 2100, drier winters, increases in extreme rainfall events during spring and summer, higher sea levels and wind speeds and changes in sediment fluxes from rivers and along the coast. Further increases in urbanisation in the region have also brought with them the potential for increased exposure to disasters, particularly of vulnerable people in informal peri-urban areas. Substantial agricultural development and intensification also raise questions about declines in the buffering capacity usually provided by healthy ecosystems and their services.

The current impacts of these extreme weather events are already evident in the large public and private sector losses in the region. Between 2003 and 2008, the Eden District accounted for 70% of the provincial government's direct disaster damage costs – US$160 million – excluding indirect damages and damages incurred by the private sector. Natural hazard claims incurred by just one short-term insurer in the Eden District over the last 15 years amounted to some US$5.5 million, with more than 78% of these claims made after 2006. Furthermore, a large national brewery identified drought as one of the major risks to their hops cultivation and therefore operations in South Africa.

Notably these impacts occur against a backdrop of large economic and social inequalities, leaving the vulnerable people and places in this region ill equipped to prepare for, cope with and adapt to disasters. The Eden District is also experiencing rapid agricultural development and urbanisation, with little information on natural hazards and options for mitigation or adaptation considered in this development.

These complex challenges sparked the initiation of the Eden Project, a co-designed and co-funded project involving research institutions, private sector partners from the insurance and beverage sectors and NGOs, together with local and provincial government agencies. These partners were motivated by increased impacts associated with natural hazards including flood, drought, wildfire and storm-waves in the region to form a collaborative research project to understand the causes of disasters and learn new ways of building resilience to disasters. Building on this growing need in Eden to address increasing risk, the aim of this use case was to integrate information on ecosystem services into land-use planning and disaster management. The concept of 'risk' was used as a boundary concept in order to bring together a diverse range of stakeholders to better understand the value of incorporating ecosystem-based management strategies into decision making, and co-design response strategies that would enhance the buffering capacity of ecosystems to mitigate the impacts of natural hazards.

Approach

The team working with a range of models, data sets as well as expert input, modelled the effect of various scenarios of climate change, as well as land cover change, on four natural hazards – floods, droughts, wildfires and storm-waves. Findings[2] showed that both climate change and land cover changes increased the frequency and intensity of natural hazards. The spread of non-native invasive trees halved the monthly river flows experienced during drought and doubled fire intensities to orders of magnitude beyond the limit for effective fire control. The impacts of plantation forestry management on floods reduced the return time between large flood events by nearly 20%. Under scenarios of moderate human-induced coastal hardening, predominantly from removal of coastal foredunes and coastal infrastructure development, severe 1:100 year storm-waves could in the future occur on an annual basis. The findings showed that through proactive land use management, it is possible to reduce the impacts of natural hazards, in some cases substantially.

These findings were used to mainstream ecosystem-based management approaches into disaster risk management through a number of products co-developed by the team of stakeholders. Product included communication presentations, brochures, papers, reports, seminars, training materials and guidelines for use in local government planning. Mainstreaming activities were guided by a systemic risk management strategy for Eden, which linked each natural hazard and its key drivers to the main institutions or communities responsible for, or with influence over, these drivers. The concept of 'risk' was thus used as a defining or boundary concept in order to bring together a diverse range of stakeholders to better understand the value of incorporating ecosystem-based management strategies into decision making, and co-design response strategies that would enhance the buffering capacity of ecosystems to mitigate the impacts of natural hazards.

Achievements

Using these products and the risk management strategy the team moved forward to identify necessary actions to reduce disaster risk in the region (Figure 3.1). Several responses, including clear outcomes and implementing agencies, were identified for each hazard. These included the development of decision support systems, inputs into management and planning processes, and the clearing and restoration of areas invaded by non-native trees or degraded by land-use practices. Partnerships were identified to address the gaps in capacity, data and resources. Partners included researchers with data and expertise, as well as the private sector and NGOs for funding, support and training.

The spread of non-native invasive trees was identified as a priority driver of vulnerability of flood, wildfire and drought risks, and resulted in a number of responses for prioritising areas and clearing non-native trees with partners from the private sector and government conservation agencies. A project to clear invasive alien trees on hops farms in Eden has now been initiated in

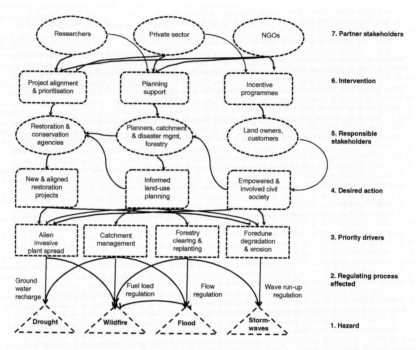

Figure 3.1 Systemic risk management strategy for Eden District of identified interventions and responsible stakeholders and actions to manage drivers of risk in the landscape.

Source: author's own

the area, which will see the investment of almost US$4 million in the area by the beverage sector, farmers, and government funding for poverty alleviation. It is estimated that uncontrolled spread of invasive non-native trees on these farms, which incur regular droughts, could reduce mean annual run-off by a massive 41% by 2032. Apart from averting the loss of much-needed water, these operations will reduce the risk of wildfire, and create in region of 200,000 employment days, or around 100 full-time jobs per year. At an average of five dependents per employee, this intervention could benefit close to 900 people, making a substantial positive contribution to the nearby small rural town with only 12,000 residents.

Similarly, a multi-million-dollar investment by South Africa's national parks was initiated to clear non-native invasive plants in and around the Garden Route National Park, a protected area embedded in the urban matrix of the Southern Cape region. Other responses include the local Fire Protection Agency now including both members from public and private sector agencies to implement and manage strategic fire belts and fuel reduction strategies. Based on findings indicating high risk of coastal storms, the implementation of foredune restoration projects has begun, along with the process to decentralize insurance underwriters to build partnerships with local authorities. At national scales the work supported inputs into national disaster management legislation and budgeting to allow for proactive ecosystem-based management for extreme events. Working with the National Disaster Management Centre and the National Treasury helped to unlock national-to-local funding barriers that exist for mainstreaming ecosystem-based management approaches into disaster risk management. Currently, local municipalities receive funding for responding to disasters only – budget allocation does not extend to investing in the longer term efforts of ecosystem-based adaptation and social learning. Inputs into the national Disaster Management Amendment Bill support this process by providing the policy support for including ecosystem-based interventions in the disaster risk management cycle.

At a local level, the material from the Eden project was used to develop brochures, decision-support toolkits, maps and annotated slide shows. Other activities focus on building networks of partners to collaborate, build capacity and coordinate activities in the region. The Eden Disaster Resilience Learning Network coordinates ecosystem-based management interventions in the area. To ensure post-project sustainability, the provincial Department of Environmental Affairs and Development Planning has agreed to chair this learning network. The learning network comprises provincial, district and local authorities from environment and disaster risk reduction, national and provincial conservation authorities, non-governmental organizations, scientists, and corporate partners in the insurance and beverage sectors. Its activities focus on co-developing and co-implementing key risk management projects in Eden in order to build the resilience of this area to climate change. The Eden Disaster Resilience Learning Network is currently establishing a data portal, on South Africa's Risk and Vulnerability Atlas website, for sharing data and information with local authorities, businesses, and beyond.

While most of the responses focused on public and private sector stakeholders, civil society was also identified as potentially powerful, including landowners and users, customers of the insurance and beverage industries and inhabitants of these areas represented through various community fora and through the media. This led to a set of responses that developed incentive programs with the insurance and beverage sector to leverage civil society influence in shaping and monitoring state planning processes, as well as in supporting restoration activities on their land.

Members of the team have supported the Business-Adopt-A-Municipality (BAAM) through board membership. The BAAM is a forum set up by the insurance sector for businesses to support local authorities in managing their risk and infrastructure. The BAAM is being piloted in Eden, but with the view to rolling out to other high-risk municipalities over time. It is supported by the South African Local Government Association, which is representative of local government and interfaces with parliament, the National Council of Provinces, cabinet as well as provincial legislatures.

CASE STUDY 2: INTEGRATED WATER RESOURCE PLANNING IN THE OLIFANTS CATCHMENT – INTEGRATING FRESHWATER ECOSYSTEMS INTO WATER RESOURCE MANAGEMENT

Introduction

The Grasslands Biome is the largest of South Africa's nine biomes, covering nearly a third of the country. The grasslands are critical water production landscapes, containing nearly half of the country's Strategic Water Source Areas. These biodiversity-rich ecosystems and their supply of ecosystem services are under tremendous pressure from booming urban development, commercial agriculture, plantation forestry and mining.

The Olifants catchment exemplifies the pressures of these grassland landscapes and is home to some of the hardest working rivers in South Africa. The catchment supports extensive coal mining and coal-fired power generation, with six of the world's largest coal-fired power stations located in the upper Olifants catchment area. The upper reaches of the catchment support 48% of the country's total power generating capacity for export and domestic consumption, includes major urban centres and steel manufacturing industries and supplies water to the second largest irrigation scheme in the country. The middle regions of the catchment are dominated by dense rural settlements, subsistence agriculture and some isolated commercial agriculture. Activities in these upper and middle reaches severely impact on the lower catchment, which flows through the country's flagship protected area: the Kruger National Park, before exiting to the sea in Mozambique. Like many other catchments in South Africa, sustainable water resource management that seeks an equitable balance between these multiple pressures on water resources is urgently required.

Capitalising on this sense of urgency, the aim of this use case was to integrate ecosystem services into water resource planning and decision-making to promote the sustainable use of water resources in the Olifants catchment. The concept of ecological infrastructure[3] was used in the Olifants catchment as a way to position ecosystem service concepts within the development priorities of South Africa, where the focus is predominantly on infrastructure. Ecological infrastructure can be seen as the nature-based equivalent of built infrastructure important for providing services and underpinning social and economic development.

Approach

A substantial portion of freshwater ecological infrastructure has recently been mapped across South Africa as Freshwater Ecosystem Priority Areas, or 'FEPA maps' (http://bgis.sanbi.org/nfepa/NFEPAmap.asp) and supporting information providing consensus on how many rivers, wetlands and estuaries, and which ones, are needed for protecting representative diversity and ecological functioning of South Africa's water resources (Nel et al. 2016). These maps were used as the knowledge product to integrate ecosystem services into water management decisions. Nel et al. (2016) also provide a review of 33 policies, and from this a policy targeting catchment-level water resource planning (known as 'Classification of Water Resources') was identified as the target decision-context for this case study. A second policy, the National Water Resource Strategy, was identified as an important enabler for incorporating ecosystem services into catchment-level policy. Department of Water Affairs government officials including the chief director, director and the project coordinator and the technical consultants were identified as key stakeholders for the use case.

The Classification of Water Resources process sets a Management Class for every significant water resource in a catchment (e.g. stretches of river). The Management Class stipulates a desired condition of the resource and the extent to which it can be utilised. The Class is determined through a scenario planning exercise and is driven by a seven-step, legislated process that provides guidance on stakeholder engagement, environmental flow assessment and scenario development. This use case explored ways to include FEPA maps, together with other technical tools such as environmental flow assessment, scenario planning and valuation, into the legislated classification process.

Achievements

The FEPA maps highlight 49 priority rivers in the Olifants catchment, which were incorporated into the scenarios. During the stakeholder negotiation process, 82% of the priority rivers were selected to be maintained in a natural or near-natural ecological condition. This decision was then presented to the minister for consideration and gazetting. The resultant method used the legislated seven-step framework guiding the process, identifying places where FEPAs need to be considered and suggesting how they can be considered. The approach was presented as a generic framework for future classification processes in other areas (Table 3.1).

Table 3.1 Process to include Freshwater Ecosystem Priority Areas in the co-designed seven-step process that guides water resource classification in South Africa

Step	Description
1. Delineate the catchment	Integrated units of analysis and significant water resources delineated. Hydrological sub-units and nodes are established as points at which ecological water requirements will be determined or extrapolated. *Allocate river nodes to FEPAs and other ecological infrastructure.*
2. Link economic and social value to ecosystem condition and water use	Develop relationships of conditions in water yield and ecosystem services to social well-being, as well as to sector-specific use and commodity generation.
3. Quantify the ecological water requirements at each node	Specify requirements that are needed to maintain each river node in a condition using environmental flow assessment approaches. *Ensure requirements for maintaining natural conditions for river nodes associated with FEPAs and any additional areas identified in Step 1.*
4. Set baseline catchment configuration	Using the requirements from Step 3, the catchment configuration with the lowest ecological condition for rivers and estuaries is established. *Explore whether it is possible to 'hardwire' the required ecological condition for river nodes FEPAs.*
5. Construct scenarios and evaluate implications	Explore scenarios including a baseline catchment configuration in Step 4, and configurations that provide for improved levels of ecological condition based on water user needs and economic development strategies. *At least two FEPA scenarios should be considered: achieving recommended ecological categories for FEPAs (a) with all other river nodes achieving the present ecological state and (b) with all other river nodes achieving the recommended ecological category. Apply a penalty to scenarios that do not achieve the recommended categories for FEPAs.*
6. Run a series of stakeholder workshops to evaluate scenarios	Set up an iterative process to evaluate the scenario outputs, and refinements of outputs, in Step 5 with quantified implications for regional economic development, social well-being and ecological conditions to assess trade-offs.
7. Select preferred scenario, assign corresponding management classes and the ecological conditions at nodes to achieve these	Management classes are assigned to Integrated Water Management Units established in Step 1 and are legally binding when they are published in the Government Gazette. Nodes, also established in Step 1 representing smaller sub-catchments within the Integrated Water Management Units, are assigned ecological condition.

Developing and documenting a mainstreaming framework

The initial draft of the mainstreaming framework was developed based on a review of mainstreaming concepts in the literature, and reflections of their application in three case studies across different use contexts and spatial scales (Sitas et al. 2014). Reflections of the case study successes were published in the peer-reviewed literature during this time, contributing to the theory and tools for mainstreaming biodiversity and ecosystem services into water resource management (Nel et al. 2017), biodiversity conservation (Manuel et al. 2016; Nel et al. 2014), co-designing research projects (Turner et al. 2016) and disaster risk reduction (Reyers et al. 2015; Sitas et al. 2016).

In reviewing the literature, the following concepts provided a very useful frame for distilling generic mainstreaming lessons across case studies, at a range of spatial scales (national, catchment, local):

Knowledge co-production: Defined as 'the collaborative process of bringing a plurality of knowledge sources and types together to address a defined problem and build an integrated or systems-oriented understanding of that problem' (Armitage et al. 2011). Unlike the traditional view of knowledge being 'produced' by researchers and then 'transferred' to users, this requires a more interactive, multi-dimensional mode of iterative knowledge co-production in a participatory arena that puts researchers, decision makers and other users of knowledge on equal footing. The process is designed to work interactively and iteratively towards collaborative learning, shared understanding of key concepts and co-evolution of common purpose, intent and action.

Legitimacy, credibility and saliency of knowledge: In order to effectively mainstream biodiversity and ecosystem services into policy and practice, knowledge co-production approaches need to pay attention to issues of legitimacy, credibility and saliency (Cash et al. 2003). The credibility of knowledge co-production is concerned with issues related to how authoritative, believable and trusted the information is. Legitimacy relates to whether the knowledge co-production process has acknowledged the diverse perspectives, values and concerns of different stakeholders. Saliency refers to how relevant the knowledge is to decision-making processes and end-user needs. Navigating the trade-offs between legitimacy, credibility and saliency are inevitable when bringing diverse knowledge and perspectives together, and this can be assisted by boundary work.

Boundary work: Navigating the interface of diverse knowledge boundaries, with stakeholders that have different perceptions on what constitutes reliable and useful knowledge can be challenging. Boundary work has been suggested as a means of managing such tensions by creating permeable knowledge boundaries, while guarding the functional integrity of contributing knowledge systems (Clark et al. 2016; Mollinga 2008). Boundary work provides three insights to use as 'tools' in this process:

- The use of suitable '*boundary concepts*', which build a common under-standing among diverse stakeholders by facilitating conceptual com-munication about the multi-dimensional nature of ecosystem services. For example, we found concepts such as 'ecological infrastructure' resonated well with stakeholders involved in infrastructure and develop-ment planning, while 'risk' was a very useful boundary concept among those involved in disaster risk management.
- The development of suitable '*boundary objects*', which establish shared approaches or methods to facilitate cooperative action among stakehold-ers. Maps, models and guidelines proved very useful in our mainstreaming work, offering objects of mutual relevance that could be applied towards the same common purpose in many different knowledge domains.
- The design of enabling '*boundary settings*' in which these concepts and methods can be developed and implemented. In the case studies, the knowledge co-production process was used an initial step towards creating this implementation-enabling environment. Finding an insti-tutional 'home' that can provide support and host a community of practice was found to be really important and was a strategy applied across case studies with effective mainstreaming outcomes.

Transdisciplinarity and the role of knowledge brokers and bridging agents: Knowledge co-production is transdisciplinary in nature, facili-tating the exchange and co-production of knowledge, not only between scientific disciplines (multi- and interdisciplinary research) but also between science and stakeholders from non-scientific disciplines (trans-disciplinary research). Such an engaged approach to research helps to uncover complementarities and create synergies across diverse knowl-edge systems. Two groups of participants appear to be vital in facilitat-ing effective knowledge exchange. The first of these, the *knowledge brokers*, comprise leaders who are able to absorb, communicate and translate knowledge across knowledge boundaries. The second group are referred to as *bridging agents*, who are able to interface effectively between the research and policy context, and are skilled at mobilising resources required for collaboration on issues of common interest, creat-ing arenas for inter-organisational learning, trust building and resolving conflict. Knowledge brokers and bridging agents can comprise individu-als, teams or organisations, are perceived to be neutral, and are trusted by the relevant parties (Reyers et al. 2015).

Proposing a mainstreaming framework and guiding principles

Drawing on these concepts, a framework was proposed to guide future main-streaming efforts (Figure 3.2), which was further refined through workshops with a broader set of scientists and practitioners working in case studies at the

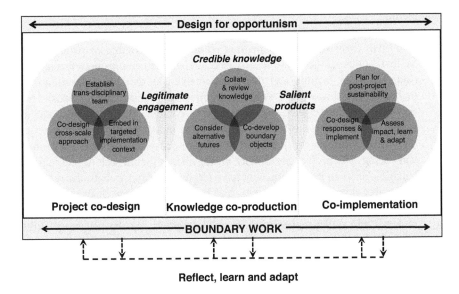

Figure 3.2 Framework for guiding mainstreaming of biodiversity and ecosystem services into policy and practice.

Source: author's own

science-policy interface. The approach comprises three knowledge exchange phases, which iteratively bring together a range of individuals, institutions, knowledge systems, skills, implementation contexts and mandates to (1) co-design the project, (2) co-produce the knowledge and outputs, and (3) co-implement, learn and adapt/transform.

Principles guiding the mainstreaming framework

There is an increasing amount of available literature on distilling lessons for mainstreaming ecosystem services into policy and practice (Albert et al. 2014; Cowling et al. 2008; Guerry et al. 2015; Maes et al. 2013; Manuel et al. 2016; Nahlik et al. 2012; Reyers et al. 2015; Sitas et al. 2014, 2016; Tallis 2011). This framework particularly highlights the importance and value of using a deeply collaborative approach which builds in cross-scale perspectives and linkages. The cross-scale perspective adopted by this framework provided much deeper insight into both the social and ecological linkages and feedbacks required for managing ecosystem services. This improved understanding helped to inform the response strategies, products and tools developed, and ultimately greatly supports uptake and impact. In the framework the cross-scale perspective was further embedded in a deeply collaborative approach, which is often referred to as knowledge co-production approaches. To support the framework, several working principles within each phase – based on key lessons across case studies – were also distilled, and include:

1   **Establish transdisciplinary project teams for knowledge co-production:** Mainstreaming projects are ideally executed as partnerships between agents (individuals or institutions) that are perceived as credible representatives of the diverse knowledge systems required for mainstreaming ecosystem services into the targeted decision making context. This is because single agents seldom have the full set of skills and mandates required to execute the transdisciplinary work needed to manage ecosystem services for human well-being. The project governance arrangements should seek to secure partnerships between institutions that have strong mandates in both the research and policy domains of ecosystems and human well-being. Such partnership enables the assembling of a transdisciplinary project team representing both research and policy negotiation, as well as providing the necessary skills and mandate for facilitating, supporting and communicating mainstreaming efforts. In a developing country context, the role of knowledge brokers and bridging agents outside of government is crucial for maintaining momentum within government, which has severe capacity constraints and is subject to high staff turnover and frequent political instability. In our case studies, knowledge brokers were drawn from research organisations, NGOs or private and public sector institutions, while bridging agents were most often drawn from NGOs or parastatals, and were chosen for their ability to play an important role in ensuring post-project sustainability.

2   **Co-design a cross-scale approach:** The approach should be co-designed by scientists, practitioners and research funders from key agencies involved in use of the co-produced knowledge. Clear aims and objectives should be articulated, along with indicators to measure outputs and outcomes. Case study reflection also highlighted the need for a cross-scale approach since local end-user perceptions and solutions are as necessary as national level policy and planning processes in order to mainstream biodiversity and ecosystem services into policy and implement ecosystem management interventions at a local level. A combination of short-term, local-scale interventions can make relatively quick progress at local levels to change behaviour, practices and management. However, these local scale interventions are greatly supported by longer-term, systemic interventions that create a more enabling implementation environment, e.g. sectoral or national policy reform. A cross-scale approach with a combination of local and national interventions has the potential to both enable context-specific local action and incorporation into higher level policy (to catalyse replication elsewhere).

3   **Embed activities in a targeted implementation context:** It is widely accepted that integrating environmental knowledge into the policy, planning, decision-making or management processes in different sectors requires an intimate understanding of the policy and institutional context in those sectors. Based on this insight, environmental projects and programmes often begin with a

legal, policy and institutional assessment. While it is possible to conduct this sort of assessment based on existing legal documentation and institutional policies and strategies, we found that the most useful information extended much further than merely knowing what policies exist and which institutions are mandated to implement them. The most useful information is not written down, and often exists only as tacit knowledge (i.e. not easily articulated). It can only be developed through substantial contact, careful listening about politics, procedures and day-to-day challenges of implementation and the nurturing of science-policy interrelationships. We found that the existing long-term networks and institutional experience of the project team members were invaluable in identifying which opportunities to target and which stakeholders to involve, thereby greatly assisting with the co-design of the project. Once opportunities have been identified, targeting specific policy opportunities enabled a much deeper form of engagement with focussed stakeholder groups, building on existing interrelationships from project team members already embedded within the implementation networks. Detailed knowledge of the decision-making tools, along with developing long-term relationships with the users of these tools, greatly facilitates the tailoring of outputs that align with end-user needs.

4   **Collate and review existing knowledge:** The aims and objectives should inform what knowledge and data are most useful. Legitimate engagement requires incorporating diverse perspectives, values and concerns of all stakeholders in the co-produced information. Striking the right balance between incorporating all stakeholder needs and perspectives and ensuring that the project's aims and objectives are achieved is hugely challenging. We found that identifying and targeting a very specific decision-making process early on in the process helped to identify the key stakeholders. Strategically targeting institutional champions and engaging with end-users to identify key decision support tools and policies perceived to influence decision making are effective starting points. To ensure credibility, we have found that it is useful to organise task teams of specialists to collate and review of this knowledge prior to its use in integrative workshops across knowledge systems. This is similar to the Multiple Evidence Base approach (Tengö et al. 2014) in which specialists add credibility to the knowledge through review within their knowledge domains. Combining credible science with political credibility (reflecting stakeholder perceptions as well as what can realistically be done by decision makers) facilitates the co-production of knowledge that is not only scientifically robust, but legitimate and aligned with end-user needs. Technical tools on their own are often insufficient to integrate into decision-making processes. Likewise, social and policy engagement can be dramatically strengthened through the use of science-based tools (such as maps and guidelines), which provide a tangible and credible basis for building a conversation with another sector. We found that using data sets endorsed by a particular sector, mandatory policy process or technical tool considerably improved in the perceived legitimacy of the resulting product, thereby supporting buy-in.

5   **Consider alternative futures for the social-ecological system:** Scenarios are plausible explorations of the future and are particularly useful in working

with stakeholders to foster a joint understanding of the interacting social and ecological features, drivers, and consequences of changes. Participants create a conceptual model of the social-ecological system and its vulnerability to change, identifying priority drivers of change and highlighting stakeholders responsible for managing these. Models of alternative futures can be developed based on these drivers of change, which are helpful in navigating the complexity of information and developing a consensus understanding of the contribution of biodiversity and ecosystem services. The alternative futures generated by this modelling process can be collectively assessed and used to inform co-designed response strategies.

6   **Co-develop boundary objects:** Boundary objects, such as standards, models and maps, establish a shared context across knowledge systems. Although they may be interpreted differently from the different sides of a knowledge boundary, they are objects of mutual interest and relevance that facilitate the negotiation and exchange of multiple types of knowledge and action. As such, the creation and management of boundary objects is a key process in developing and maintaining coherence across many different knowledge domains. We have found common terminology (e.g. 'ecological infrastructure', 'risk', 'resilience'), maps of priority areas for action accompanied by implementation guidelines and conceptual models are important boundary concepts that can catalyse different actions across sectors that all work towards a common purpose (Case Study 1).

7   **Co-design responses and implement:** This step involves co-designing necessary actions and outcomes with specific stakeholders responsible for managing the drivers of change and implementing the actions. In designing response strategies, we have found that it is important to upscale response options from local to national levels, as well as to upscale sector-wide response strategies (as opposed to single agency). For example, engagement with both local and national decision makers can result in uptake of the guidelines not only in the local area of concern, but also in similar work being undertaken elsewhere in the country. Similarly, building in a sector-wide response allows for far greater impact beyond a single agency in places elsewhere (e.g. Case Study 2). Boundary objects can also greatly assist with this upscaling.

8   **Plan for post-project sustainability:** By paying attention to legitimacy, credibility and saliency, a foundation will have already been built for implementation. However, the lifetime of one mainstreaming project is typically shorter than the 7–10 years required for embedding institutional change. It is therefore important to plan for post-project support and build the capacity of young scientists and practitioners, as well as establish and nurture learning networks throughout the project in order for work to continue beyond the project's timeframe. Here, the role of bridging organisations in facilitating future collaboration and knowledge co-production among stakeholder groups is critical. Planning for post-project sustainability also includes having multiple implementation strategies and responsibilities tailored for each organisation (see no. 9 below), and writing joint proposals for future work

that can assist with sustaining mainstreaming efforts. Continuity from the start of the project process through to actual on-ground implementation is ideal, and therefore it is important to identify and secure the involvement of potential stakeholders that can serve as bridging organisations upon the completion of the project.

9   **Design for opportunism:** Taking advantage of windows of opportunity that arise is another widely reported precondition for mainstreaming success. These opportunities might include actively pursuing negotiation around policies or plans that are under development, revision or amendment (e.g. national development plans). Having a broad range of potential communication materials (e.g. maps on ecosystem services, tools, statistics and infographics or guideline documents) can greatly enhance the potential for quick reaction should a window of opportunity arise. Operationally within projects, fast reaction to windows of opportunity presents some challenges, as it requires flexibility both in the work promised and the network of experts that may be required. While there are benefits to tightly managed projects with logframes, this is often the antithesis of the flexibility that is required, and such uncertainty can be unsettling for funders. A balance between tight project management and flexibility to harness opportunities can be built in through processes like a formal mid-term review which allows sufficient time for the project team and stakeholders to mature their ideas, and make refinements that would focus the work and have greatest likelihood for mainstreaming impact based on new opportunities that had developed since project initiation.

Assessing mainstreaming impacts

In mainstreaming ecosystem services into decision and policy making it is useful to reflect on what is meant by successful mainstreaming. While the ultimate objective of mainstreaming is policy change and a resultant positive impact on ecosystem services and human well-being, this is often not possible within the usual time frame of projects. Instead is it helpful to measure progress along this route to ultimate impact using measures such new knowledge jointly produced or changes in awareness or understanding (Figure 3.3). These earlier stages of impact are important precursors to ultimate mainstreaming, paving the way for when a policy engagement window opens and sometimes even helping the window to open. Furthermore, they are important measures of impact themselves, changing the way that issues are often framed or explored and introducing new options for consideration in policy and decision contexts.

Often there is ample evidence of the first stage of mainstreaming success – joint knowledge production. New databases, reports, maps, models, guidelines, papers and presentations are not only evidence of new knowledge, but of co-authorship and co-ownership of this knowledge. The framework for water resource classification developed in Table 3.1 is an example of successfully engaged stakeholders in the co-production of knowledge on ecosystem services. These co-produced outputs represent important resources tailored for use in specific decision contexts,

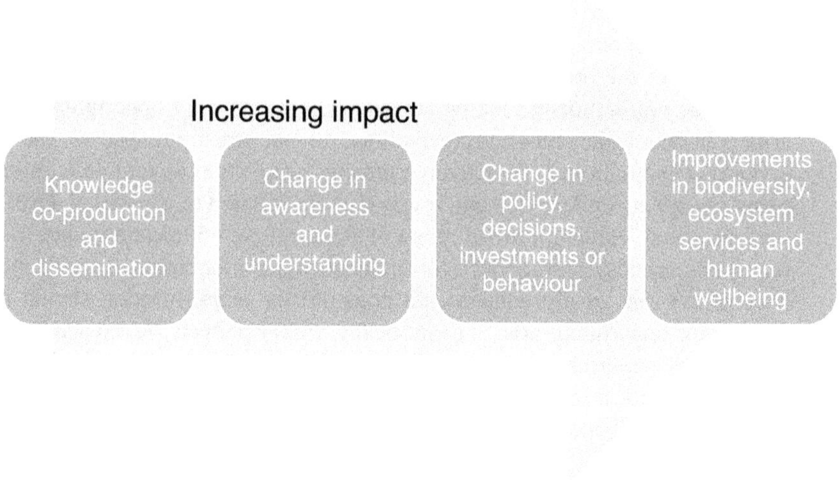

Increasing impact

Knowledge co-production and dissemination

Change in awareness and understanding

Change in policy, decisions, investments or behaviour

Improvements in biodiversity, ecosystem services and human wellbeing

Figure 3.3 Levels of impact and milestones of success in mainstreaming ecosystem services into policy and practice.

Source: adapted from Ruckelshaus et al. 2015

and will also have secured support for their use in these contexts through the involvement of high level officials in their production. They therefore will support mainstreaming into policy when opportunities for policy or decision change present themselves. Furthermore, they contribute to the next stage of mainstreaming success through creating new understanding and awareness of ecosystem services across stakeholder groups, shaping and influencing the perceptions of the importance of ecosystems and their services.

Measures of success in shaping awareness can also be assessed e.g. through the measurable change in the levels of understanding about ecosystem services in stakeholders from across the public and private sectors linked to communications materials and outreach activities. The toolkits and materials developed, together with training on their use, should serve to further this impact on awareness of the relevance of biodiversity and ecosystem services. Further signs of changes in awareness and understanding can also be found where previously unconnected stakeholder groups from the private and public sector interact around the topic of ecosystem-based management. This shift in how the issue is framed – from a focus on climate, infrastructure or disaster response to one on ecosystem management – is a clear indication of a shift in understanding on the role of ecosystem services (Case Study 1; Reyers et al. 2015). Similarly, the new links with policy makers from the non-environment sectors can signal a shift in the understanding of the links between ecosystems and broader development issues including disaster management, development planning and infrastructure planning (Case Study 2). The proposal for an additional investment in large program on restoring ecosystems for water security seen in South Africa is a strong signal of a shift in understanding of the relevance of ecosystems to national development targets. Also seen are new levels of awareness of the links between ecosystems and business interests from areas as diverse as wine to insurance (see Reyers et al. 2015; Sitas et al. 2016).

In examining measures of impact beyond awareness to policy change, early examples in South Africa include some policy changes including an expansion to include ecosystem services in biodiversity planning policy and the inclusion of water ecosystems and services in water resource management policy. Similarly, solicited comments and inputs into disaster management policy and budgeting could lead to large shifts in policy in these areas. Continued support for these policy processes is thus an important component of projects, which require careful design of post-project sustainability and continued engagement with policy beyond the project's end.

Beyond policy change there could be some signs of impacts on the ground with possible positive outcomes for biodiversity and ecosystem services. Establishing monitoring programs in these cases forms an important component of post-project sustainability through learning networks, the handover of project outputs and monitoring requirements to implementing agencies and the links to policies with monitoring requirements.

Conclusion

Ecosystem-service approaches offer the potential to create change in the perceptions, investments, decisions and policies needed to achieve improved outcomes for ecosystems, human well-being and sustainable development. However, due to gaps in the necessary scientific knowledge and tools, as well as the transdisciplinary challenges associated with the complex social and political processes needed to catalyse action, this potential remains unrealised. A possible solution to some of these challenges may lie in the practice of ecosystem assessments. Reflecting across these projects, it is clear that multiple methods, approaches and forms of engagement are required for mainstreaming ecosystem service knowledge into decision making.

Notes

1 Global Assessment Report on Disaster Risk Reduction 2013: www.preventionweb.net/english/hyogo/gar/2013/en/home/index.html
2 Reyers, B., Nel, J.L., O'Farrell, P.J., Sitas, N. and Nel, D.C., 2015. Navigating complexity through knowledge coproduction: Mainstreaming ecosystem services into disaster risk reduction. *Proceedings of the National Academy of Sciences*; Nel, J.L., Le Maitre, D.C., Nel, D.C., Reyers, B., Archibald, S., van Wilgen, B.W., . . . Barwell, L. (2014). Natural hazards in a changing world: A case for ecosystem-based management. *PloS One*, 9(5).
3 Ecological infrastructure refers to functioning ecosystems that deliver valuable services to people, such as clean water, climate regulation, soil formation and disaster risk reduction.

References

Albert, C., Aronson, J., Fürst, C. and Opdam, P., 2014. Integrating ecosystem services in landscape planning: Requirements, approaches, and impacts. *Landscape Ecology, 29*(8), pp. 1277–1285.

Armitage, D., Berkes, F., Dale, A., Kocho-Schellenberg, E. and Patton, E., 2011. Co-management and the co-production of knowledge: Learning to adapt in Canada's Arctic. *Global Environmental Change, 21*(3), pp. 995–1004.

Biggs, R., Bohensky, E., Desanker, P.V., Fabricius, C., Lynam, T., Misselhorn, A.A., Musvoto, C., Mutale, M., Reyers, B., Scholes, R.J., Shikongo, S., van Jaarsveld, A.S., 2004. *Nature Supporting People: The Southern African Millennium Ecosystem Assessment.* CSIR, Pretoria.

Bohensky, E. and Lynam, T., 2005. Evaluating responses in complex adaptive systems: Insights on water management from the Southern African Millennium Ecosystem Assessment (SAfMA). *Ecology and Society, 10*(1), pp. 1–153.

Bohensky, E., Reyers, B., Van Jaarsveld, A.S. and Fabricius, C., 2004. Ecosystem services in the Gariep basin: A basin-scale component of the Southern African Millennium Ecosystem Assessment (SAfMA). *African Sun Media*, Stellenbosch, South Africa.

Cash, D.W., Clark, W.C., Alcock, F., Dickson, N.M., Eckley, N., Guston, D.H., Jäger, J. and Mitchell, R.B., 2003. Knowledge systems for sustainable development. *Proceedings of the National Academy of Sciences, 100*(14), pp. 8086–8091.

Clark, W.C., Van Kerkhoff, L., Lebel, L. and Gallopin, G.C., 2016. Crafting usable knowledge for sustainable development. *Proceedings of the National Academy of Sciences, 113*(17), pp. 4570–4578.

Cowling, R.M., Egoh, B., Knight, A.T., O'Farrell, P.J., Reyers, B., Rouget, M., Roux, D.J., Welz, A. and Wilhelm-Rechman, A., 2008. An operational model for mainstreaming ecosystem services for implementation. *Proceedings of the National Academy of Sciences, 105*(28), pp. 9483–9488.

Dalal-Clayton, D. B. and Bass, S., 2009. *The Challenges of Environmental Mainstreaming: Experience of Integrating Environment into Development Institutions and Decisions.* IIED, London.

Díaz, S., et al. 2015. The IPBES conceptual framework: Connecting nature and people. *Current Opinion in Environmental Sustainability, 14*, pp. 1–16.

Guerry, A.D., Polasky, S., Lubchenco, J., Chaplin-Kramer, R., Daily, G.C., Griffin, R., Ruckelshaus, M., Bateman, I.J., Duraiappah, A., Elmqvist, T. and Feldman, M.W., 2015. Natural capital and ecosystem services informing decisions: From promise to practice. *Proceedings of the National Academy of Sciences, 112*(24), pp. 7348–7355.

Maes, J., Hauck, J., Paracchini, M.L., Ratamäki, O., Hutchins, M., Termansen, M., Furman, E., Perez-Soba, M., Braat, L. and Bidoglio, G., 2013. Mainstreaming ecosystem services into EU policy. *Current Opinion in Environmental Sustainability, 5*(1), pp. 128–134.

Manuel, J., Maze, K., Driver, M., Stephens, A., Botts, E., Parker, A., Tau, M., Dini, J., Holness, S. and Nel, J., 2016. *Key Ingredients, Challenges and Lessons from Biodiversity Mainstreaming in South Africa.* OECD Environment Working Papers, No. 107, OECD Publishing, Paris.

Millennium Ecosystem Assessment (MA), 2005. *Millennium Ecosystem Assessment: Ecosystems and Human Well-Being.* Synthesis, Washington, DC.

Mollinga, P.P., 2008. *The Rational Organisation of Dissent: Boundary Concepts, Boundary Objects and Boundary Settings in the Interdisciplinary Study of Natural Resources Management* (No. 33). ZEF Working Paper Series. Department of Political and Cultural Change Center for Development Research, University of Bonn, Bonn.

Nahlik, A.M., Kentula, M.E., Fennessy, M.S. and Landers, D.H., 2012. Where is the consensus? A proposed foundation for moving ecosystem service concepts into practice. *Ecological Economics, 77*, pp. 27–35.

Nel, J.L., Le Maitre, D.C., Nel, D.C., Reyers, B., Archibald, S., van Wilgen, B. W., Forsyth, G.G., Theron, A.K., O'Farrell, P.J., Mwenge Kahinda, J.-M., Engelbrecht, F.A., Kapangaziwiri, E., van Niekerk, L. and Barwell, L., 2014. Natural hazards in a changing world: A case for ecosystem-based management. *PloS One, 9*(5), p. e95942.

Nel, J.L., Le Maitre, D.C., Roux, D.J., Colvin, C., Smith, J.S., Smith-Adao, L.B., Maherry, A. and Sitas, N., 2017. Strategic water source areas for urban water security: Making the connection between protecting ecosystems and benefiting from their services. *Ecosystem Services, 28*, pp. 251–259.

Nel, J.L., Roux, D.J., Driver, A., Hill, L., Maherry, A.C., Snaddon, K., Petersen, C.R., Smith-Adao, L.B., Van Deventer, H. and Reyers, B., 2016. Knowledge co-production and boundary work to promote implementation of conservation plans. *Conservation Biology, 30*(1), pp. 176–188.

Reyers, B., Nel, J.L., O'Farrell, P.J., Sitas, N. and Nel, D.C., 2015. Navigating complexity through knowledge coproduction: Mainstreaming ecosystem services into disaster risk reduction. *Proceedings of the National Academy of Sciences*, p. 201414374.

Ruckelshaus, M., McKenzie, E., Tallis, H., Guerry, A., Daily, G., Kareiva, P., Polasky, S., Ricketts, T., Bhagabati, N., Wood, S.A. and Bernhardt, J., 2015. Notes from the field: Lessons learned from using ecosystem service approaches to inform real-world decisions. *Ecological Economics, 115*, pp. 11–21.

Shackleton, C., Fabricius, C., Ainslie, A., Cundill, G., Hendricks, H., Matela, S. and Mhlanga, N., 2004. *Southern African millennium assessment: Gariep Basin Local Scale Assessments*. Rhodes University, Grahamstown, South Africa.

Sitas, N., Prozesky, H.E., Esler, K.J. and Reyers, B., 2014. Opportunities and challenges for mainstreaming ecosystem services in development planning: Perspectives from a landscape level. *Landscape Ecology, 29*(8), pp. 1315–1331.

Sitas, N., Reyers, B., Cundill, G., Prozesky, H.E., Nel, J.L. and Esler, K.J., 2016. Fostering collaboration for knowledge and action in disaster management in South Africa. *Current Opinion in Environmental Sustainability, 19*, pp. 94–102.

Tallis, H., 2011. *Natural Capital: Theory and Practice of Mapping Ecosystem Services*. Oxford University Press, Vancouver.

Tengö, M., Brondizio, E.S., Elmqvist, T., Malmer, P. and Spierenburg, M., 2014. Connecting diverse knowledge systems for enhanced ecosystem governance: The multiple evidence base approach. *Ambio, 43*(5), pp. 579–591.

Turner II, B.L., Esler, K.J., Bridgewater, P., Tewksbury, J., Sitas, N., Abrahams, B., Chapin III, F.S., Chowdhury, R.R., Christie, P., Diaz, S. and Firth, P., 2016. Socio-Environmental Systems (SES) Research: What have we learned and how can we use this information in future research programs. *Current Opinion in Environmental Sustainability, 19*, pp. 160–168.

Van Jaarsveld, A.S., Biggs, R., Scholes, R.J., Bohensky, E., Reyers, B., Lynam, T., Musvoto, C. and Fabricius, C., 2005. Measuring conditions and trends in ecosystem services at multiple scales: The Southern African Millennium Ecosystem Assessment (SAfMA) experience. *Philosophical Transactions of the Royal Society of London B: Biological Sciences, 360*(1454), pp. 425–441.

4 Linking growth and natural capital in the Republic of Kazakhstan

A case study of Aral Sea

Terry Roe and Rodney Smith

Introduction

Economic growth can impact the condition of nature and ecosystems if the growth is not extracted in environmentally friendly manner. There are several examples of changes in iconic ecosystems across the world. The disappearance of the Aral Sea in central Asia is one such example which has drawn the global attention from scientists and policy makers.

Micklin (2014) notes in 1960 the Aral Sea was the fourth largest lake in the world, with a surface area of over 67,000 km^2 and a volume of almost 1,100 km^3. By 2011, its surface area had shrunk to about 12,000 km^2, and its volume fell to about 88 km^3. A major contributor to the Aral Sea's demise was the increased level of irrigation along its two tributary rivers – the Amu Darya and Syr Darya – that saw irrigated area grow from 5 million hectares in 1960 to 8.2 million hectares by 2010. The increased levels of irrigation combined with the natural evaporation levels of the Aral Sea led to a significant fall in the net flow of water entering the sea (Micklin, 2014). Over time, consistent low water inflow levels triggered significant declines in the Aral's surface area and volume.

By 1987, the Aral Sea had roughly divided into two major water bodies: the North Aral Sea and the South Aral Sea (Micklin, 2014). The North Aral lay entirely within Kazakhstan, with its water inflows deriving mostly from the Syr Darya Basin. The boundaries of the South Aral Sea lay within both Kazakhstan and Uzbekistan, with its water inflow deriving mostly from the Amu Darya Basin. By 2012, the South Aral had receded to the East-South Aral Sea, while surface area and volume increased somewhat in the North Aral Sea.

According to Micklin (2014), restoring the Aral Sea to its 1960s levels is possible, but unlikely "in the foreseeable future." Doing so would require average river inflow levels of about 56 km^3 per year, and take over 100 years; the sea could reach 91% of its former state, however, in 43 years. Increasing inflow levels to 56 km^3 per year is unlikely, but regional efforts to restore the North Aral Sea have been quite successful. To raise water levels, increase surface area and lower salinity levels in the North Aral Sea, local agencies built an earthen dike in 1992 to control the outflow of water down the channel to the South Aral. The dike was breached several times and eventually replaced with the structurally sound Kok-Aral Dike, completed in 2005. See Micklin (2014) or Micklin, Aladin and Plotnikov (2014) for further details. The Kok-Aral Dike did, in fact, help raise the water level and decrease salinity levels, and only required 3.5 km^3 of inflow each year: in 2003, the North Aral's surface area was 3,200 m^2 and 30 meters deep – by 2006 its surface area was 3,600 m^2 and 42 meters deep (Japan Aerospace Exploration Agency).

The Kazakh government is about to begin another phase of North Aral restoration, by building a dike and dam at the mouth of the Gulf of Saryshaganak, and divert water to the gulf. The major goal of this project is to raise the water level in the gulf to 50 meters, and in doing so, extend the surface area of the North Aral Sea to Aralsk, a town once famous for its active fishing industry. A recent World Bank document (Ghany and Shawky, 2014) suggests US$126 million has been earmarked for the project.

To raise the water level in the Gulf of Saryshaganak and maintain the height and salinity levels of the entire North Aral Sea requires water from the Syr Darya, which to date is mostly allocated to agriculture or North Aral Sea restoration. Table 4.1 illustrates how Syr Darya water was allocated across competing uses in Kyzylorda and South Kazakhstan, between 2006 and 2014. In both regions, water intake levels directed to agriculture accounted for the lion's share of total intakes, with Kyzylorda agriculture receiving over 99% of water, and South Kazakhstan agriculture receiving at least 96% of water. The data suggest that, on average, Kyzylorda agriculture and South Kazakhstan industry received increased water assignments over the period. Data not reported here suggest the increased intake levels for Kyzylorda agriculture is the result of investments in irrigation canal restoration and repair.

This chapter examines *part* of the economic trade-off between water uses across the two activities. Specifically, it develops a mainstreaming tool that

Table 4.1 Water intake levels in Kyzylorda and South Kazakhstan (million m^3)

	Sector	2008	2009	2010	2011	2012	2013	2014
Kyzylorda	Agriculture	3,053.5	3,429.1	3,457.0	3,632.4	3,717.9	3,563.8	3,785.9
	Industry	5.9	5.0	4.0	5.3	4.8	4.3	3.4
	Household	9.6	8.4	10.4	9.4	9.5	8.5	6.1
	Fishery	4.6	4.6	4.6	4.6	4.8	4.8	4.8
	Total	**3,073.6**	**3,447.1**	**3,476.0**	**3,651.7**	**3,737.0**	**3,581.4**	**3,800.2**
S. Kazakhstan	Agriculture	2,926.1	3,172.4	2,968.2	3,212.8	3,804.6	3,754.6	3856.9
	Households	37.5	36.9	36.5	36.3	29.8	32.4	33.3
	Industry	48.1	51.4	56.4	61.7	79.9	78.9	79.8
	Fishery	8.3	9.3	8.9	9.1	15.4	13.4	13.0
	Total	**3,020.0**	**3,270.0**	**3,070.0**	**3,319.9**	**3,929.7**	**3,879.3**	**3,983.0**

Source: CAREC, 2015

Data provided by the Regional Environmental Centre for Central Asia (CAREC), in cooperation with the Kazakhstan Ministry of Agriculture, Committee of Water Resources.

measures the economic value of water in agricultural production along the Syr Darya basin: doing so gives us a better idea of water's importance to the regional economy, and an idea of what is at state when asking farmers to divert water from agricultural uses to the North Aral Sea. The analysis that follows uses two measures of value: one measure is the shadow rental value of water and land used in agricultural production; the other is the stock – or investment – value of water. The *shadow rental value* of water (and land) is the resource's contribution to gross domestic product (GDP) – sometimes referred to as the *value added* of the resource – and is the amount farmers would be willing to pay for to use additional units of water (over a specific time period, say a growing season). The stock value of water (and land) is the amount farmers would pay to own the resource, and is calculated as the discounted present value of future shadow rental values.

We use a dynamic, general equilibrium model to measure the shadow rental values. We do this because water's shadow rental value is heavily influenced by the levels of other resources with which it is combined to produce agricultural output. For instance, if labor is a scarce resource, then combining a little more labor with water, capital, fertilizer and land increases the productivity of water (and the other factors) – meaning, using a little more labor increases the shadow rental value of water. Another reason for using such a framework is natural resource and physical capital stock values are relative. Smith, Nelson and Roe (2015) show that the stock values of water and land hinge crucially on the stock price of (manmade) capital, and that the values can move together. They also show when this connection is ignored, stock measures of land and water values in Punjab agriculture are seriously underestimated – yielding stock values less than one-third or more of their correctly measured values. Smith and Gemma (2014) find similar results for water and land used in Japanese agriculture.

One implication of the Smith, Nelson and Roe (2015) study is, unless the stock value of a natural resource is linked to the stock value of capital, the estimated stock values of the natural resource are likely to be biased – and the bias could be significant. Their results also suggest a resource's value over time is influenced by the relative competitiveness of the sector using it, where a sector's competitiveness is influenced by its relative capital intensity. Their empirical results suggest, on average, the relatively more capital-intensive sectors realize faster increases in shadow rental values over time. This observation suggests competitiveness affects the future shadow values of a resource, and hence, its value is influenced by the choices made by agents in other parts of the economy.

The empirical analysis of this study examines three policy options.

1   The status quo policy establishes the baseline income across seven productive sectors and focuses attention on the shadow rental values of land and water in agricultural production over time. We then use this data to calculate agricultural wealth – the asset value of land and water – over time.
2   Could we increase natural resource asset values by allowing oblasts to trade water among themselves? The answer is yes, with South Kazakhstan typically

renting water from the other regions along the Syr Darya, and total asset values increasing by a little over 1.6%.

3 The third policy examines the wealth impact of improving canal efficiency along the Syr Darya. The results suggest farmers improving irrigation efficiency increases the total value of land and water wealth by a nominal amount. These results, however, should be treated as preliminary, as the water use trading model's specification likely leads to an overestimate of the potential gain of the policy. On the other hand, the specification of the improved irrigation efficiency model likely leads to an underestimate of the policy's impact.

The current chapter should not be viewed as an attempt measure, fully, the trade-offs of allocating water between agriculture and Syr Darya restoration/maintenance, as we only measure the potential income farmers forgo when using less water and, say, not expand cultivated area. A full valuation would measure the economic benefit to the region of the expanded fishery industry accompanying an improved North Aral Sea ecosystem. Such an analysis, however, was beyond the scope of the current study.

The next section, "Literature Review", provides an overview of prior analysis of Syr Darya Basin economics, and the third section, "Model Basics" provides a brief description of the economic model used to conduct the valuation and policy impact analysis – details of the model are relegated to Appendix 4.1. The fourth section, "Data", describes the data used to parameterize the empirical model(s) that follow, and their sources. The fifth section "Simulation", presents the empirical results from the four policy simulations. The last section, "Conclusions", sums up results of the analysis and suggests future studies.

Literature review

A relatively large literature exists on the history of the physical characteristics of the Aral Sea and how irrigation activities along the Amu and Syr Darya led to its demise. Attention has also focused on the success of the Kazakh government's efforts to restore the North Aral Sea. See Micklin, Aladin and Plotnikov (2014) for a comprehensive review of the literature on these topics and summary of the Aral Sea's rehabilitation efforts.

One point that emerges from the Aral Sea's story is that Kazakhstan's water endowments depend critically on how much water its upstream neighbors, Tajikistan and Kyrgyzstan and Uzbekistan, release downstream: South Aral Sea restoration relies mostly on water from the Amu Darya via Tajikistan, while North Aral Sea restoration relies mostly on water from the Syr Darya via Uzbekistan and Kyrgyzstan. The literature suggests the relationship between the countries has sometimes been contentious, but the countries have evolved a mutually beneficial sharing of resources. Kazakhstan, Uzbekistan and Kyrgyzstan share energy and water resources – with Kyrgyzstan providing Kazakhstan and Uzbekistan electric power and releasing water for agricultural production, in return for coal, natural

Table 4.2 Land and water productivity values (rental values)

	Land (US$/hectare)		Water (US$/km³)	
	Kyzylorda	*S. Kazakhstan*	*Kyzylorda*	*S. Kazakhstan*
Cooperatives	552	599	30	150
Private farms	725	1,475	40	220
District water management organizations	452	591	20	140

Source: adapted from IWMI, 2003
www.iwmi.cgiar.org/Publications/IWMI_Research_Reports/PDF/Pub067/Report67.pdf

gas and oil from Kazakhstan and Uzbekistan (see Murray-Rust et al., 2003; Micklin, Aladin and Plotnikov, 2014).

To our knowledge, few published studies exist that examine water valuation or water productivity in agriculture. One notable exception is a 2004 World Bank study that suggests the value of irrigation water along the Syr Darya ranges from US$20 to $50 per thousand cubic meters (km³). Another is a 2003 study by the International Water Management Institute (IWMI), that develops estimates of land productivity (in US$/hectare) and water productivity (in US$/m³) in several regions of Kazakhstan and Uzbekistan. The study included estimates of land productivity (rental) values of cotton, wheat and rice in South Kazakhstan and in Kyzylorda. Both the land and water productivity values, however, are average values for the production of all three crops, not individual values. The IMWI land and water productivity values for Kyzylorda and SK are repeated in Table 4.2.

Although studies exist that estimate the flow value of water along the Syr Darya, this report is almost certainly the first attempt to estimate the stock value of Syr Darya water and link it to North Aral Sea recovery.

Water allocation along the Syr Darya

The supply of water to South Kazakhstan and Kyzylorda is the result of a complex allocation process. First the Syr Darya Basin Valley Organization (BVO) and Amu Darya BVO collect data on water demand from provincial water management units called oblvodkhozes and send the demand data to the Committee for Water Resources under the Ministry of Agriculture and Water Management. Next, the Committee for Water Resources decides how much water goes to the irrigation systems within each state, including interstate, interdistrict and interfarm canals. These limits are sent to BVO and oblvodkhozes, and the latter determine how much water to distribute to each district and farm on the base of agreements with farmers.

The water allocation process uses three volume measures: the irrigation water limit (IWL), irrigation water demand (IWD) and irrigation water supply (IWS). The IWL is the maximum amount a region will be allocated and is typically linked to projected water availabilities. The IWD is an estimate of the amount of water a region (province, oblast and district) would demand given its production goals, climate and soil conditions. The IWS is the actual amount of water allocated to the region, and will not necessarily be equal to a region's IWL. Again, see Murray-Rust et al. (2003) for more details. Table 4.3 presents Kyzylorda's and South Kazakhstan's IWD, IWL and IWS levels for the years 2005 through 2014.

Table 4.3 suggests South Kazakhstan's water demand exceeded its allocation each year. Anecdotal evidence suggests water supplies had fallen over time, but canal restoration efforts are beginning to increase water availability to the region. Straightforward calculations suggest that on average, of the Syr Darya water allocated to Kyzylorda, 90% of that water makes its way into the Kyzylorda canal system: of the Syr Darya water allocated to South Kazakhstan, about 78% of that water enters the South Kazakhstan canal system. Recently, Bekchanov, Bhaduri and Ringler (2015) suggest Kyzylorda and South Kazakhstan could increase canal conveyance efficiency from 70% to 90% (local experts place a cap on canal conveyance efficiency at 85%).

The conceptual model, and the corresponding empirical model, divides Kyzylorda agriculture into two sectors; rice and other-Kyzylorda agriculture. The reason for this is hinted at in Table 4.4, which reveals that rice production typically accounts for more than 80% of Kyzylorda agricultural water use. We also

Table 4.3 Average water allocations to Kyzylorda and South Kazakhstan (in m³/season)

		2007	2008	2009	2010	2011	2012	2013	2014
Kyzylorda	IWD	3,652.0	3,200.0	3,605.5	4,027.0	3,953.7	3,818.8	4,088.7	4,152.3
	IWL	3,652.0	3,200.0	3,605.5	4,027.0	3,953.7	3,818.8	4,088.7	4,152.3
	IWS	3,570.5	3,053.5	3,429.1	3,457.0	3,632.4	3,717.9	3,563.9	3,786.0
S. Kazakhstan	IWD	4,380.0	4,392.0	4,410.0	4,438.0	4,433.0	4,453.0	4,469.0	4,629.0
	IWL	3,100.0	2,659.0	2,948.0	2,649.0	2,880.0	2,794.0	3,047.0	3,267.0
	IWS	2,443.0	2,115.0	2,384.0	2,175.0	2,310.0	2,213.0	2,413.0	2,540.0

Source: CAREC, 2015

Table 4.4 Water allocation shares

	2007	2008	2009	2010	2011	2012	2013	2014
Kyzylorda								
Rice	0.820	0.806	0.809	0.806	0.818	0.769	0.779	0.831
Lucerne	0.081	0.086	0.090	0.089	0.079	0.132	0.118	0.093
Other agriculture	0.098	0.107	0.101	0.105	0.103	0.099	0.103	0.076
South Kazakhstan								
Cotton	0.471	0.410	0.303	0.288	0.312	0.291	0.256	0.234
Forage (annual and perennial grasses)	0.261	0.288	0.354	0.352	0.336	0.288	0.353	0.353
Other agriculture	0.179	0.203	0.236	0.254	0.266	0.330	0.316	0.314

Source: CAREC, 2015

disaggregate South Kazakhstan agriculture into two subsectors; cotton and other agriculture, primarily because of cotton's dominance in water use before 2007. Furthermore, available data suggest water productivity in cotton is quite different than that of water productivity in the rest of South Kazakhstan, with water accounting for 6% of non-cotton value added and accounting for 16% of cotton value added (see the social accounting matrix in Appendix 4.2).

Policy issues: As noted previously, this study examines three policy questions. The status quo policy is to keep IWL levels unchanged over time. The second policy examines the impact of keeping IWL levels unchanged but allowing oblasts to trade water use rights each year. The third policy examines the economics of keeping IWL unchanged but improving irrigation efficiency. For each scenario we project the land and water flow values over time and calculate the corresponding stock values of the two assets.

Model basics

The valuation exercise implemented in this study takes as its point of departure the theoretical framework and corresponding empirical methodology presented in Roe, Smith and Saraçoğlu (2010). Although the models developed in Roe, Smith and Saraçoğlu (RSS) are dynamic, they each have a static and a dynamic component. The static component models the behavior of two types of agents – consumers and producers – and describes the results of their interactions. Producers combine capital, labor and other inputs to produce final goods and services. Consumers use income to purchase final goods and services (today) and save (for future consumption). The groups of agents interact in "markets" which help determine how resources ultimately get allocated across competing demands. The dynamic component models the optimal savings and consumption decisions of consumers over time.

The static component begins with a utility function for the "representative consumer," and a production function for each productive sector of the economy. The utility function is used to derive an *expenditure function* for the consumer, which the theoretical and empirical model uses to determine how much of a good households demand (or how much they will spend on the good). The production function is used to derive a *cost function* or *value added function* for each sector. The theoretical and empirical model uses the cost and value-added functions to predict how much capital, labor and other inputs a sector will demand, and how much output it will produce. The economy combines physical capital, labor, land and water to produce agricultural, manufactured and service goods. Although other natural resources like minerals and oil are natural endowments, we ignore them in the analysis that follows, and bury them into the economics governing the manufacturing sector.

Capital and labor are used by each sector, and are mobile across the economy. Land is used in agricultural production only and is fixed to a region or sector. Water will initially be viewed as allocated to a specific sector, but we relax this condition when examining the water use rights trading policy options. We assume all final good markets are perfectly competitive – i.e., no single consumer or producer can influence market prices.

In the models that follow, the structure of agricultural production is such that South Kazakhstan is the major region producing cotton, while Kyzylorda is the major region producing rice. Both regions produce other agriculture, and in South Kazakhstan we aggregate non-cotton production into "SK other agriculture" and in Kyzylorda we aggregate all non-rice production into "Kyzylorda other agriculture." In the empirical model cotton receives 34% of IWS in South Kazakhstan and South Kazakhstan other agriculture receives the remaining 66%, while rice receives 81% of IWS in Kyzylorda and Kyzylorda other agriculture receives the rest.

Our current understanding is that manufacturing contributes a relatively small amount to Kyzylorda or South Kazakhstan gross domestic product (GDP). Also, manufacturing uses very little water drawn from the Syr Darya. Given these two conditions, we decided to integrate Kyzylorda and South Kazakhstan manufacturing GDP into an aggregate manufacturing sector for all of Kazakhstan. The same reasoning applied to the service sector. Hence, the conceptual (and analogous empirical) model has seven sectors: cotton, rice, other agriculture in South Kazakhstan, other agriculture in Kyzylorda, other agriculture in the rest of Kazakhstan, manufacturing and services.

Data

We parameterize the model using data from several sources. The major data source is a year 2007 social accounting matrix (SAM) for Kazakhstan provided by CAREC. The SAM was aggregated to match the seven sectors discussed previously: South Kazakhstan cotton, Kyzylorda rice, South Kazakhstan rest of agriculture, Kyzylorda rest of agriculture, the rest of Kazakhstan other agriculture, (all of Kazakhstan) manufacturing and services. The SAM factor categories include capital, labor, land and water. Table 4.5 lists the factor shares for each sector.

The second major data source is the World Bank's World Development Indicator (WDI) (World Bank, 2015) data on Kazakhstan's gross fixed capital formation, labor force and gross domestic product, with the WDI data used to create a capital stock series. The third data source is hydrological data from various Kazakh and web-based publications and CAREC experts.

The labor force, gross fixed capital formation and GDP data serve two purposes. First, they allow us to estimate a capital stock series for Kazakhstan, which

Table 4.5 Kazakhstan factor shares

	Cotton	Rice	Other Agriculture			Manufacturing	Services
			SK	Kyzylorda	ROK		
Labor	0.600	0.600	0.600	–	–	0.374	0.630
Capital	0.120	0.072	0.122	–	–	0.626	0.370
Land	0.165	0.278	0.066	–	–	–	–
Water	0.115	0.050	0.212	–	–	–	–

Source: author adjustments to Roland-Holst and Kazybayeva, 2009

Table 4.6 Kazakhstan consumption shares

Rice	Other Ag.	Industry	Services
0.00004	0.06618	0.1621	0.7716

combined with the social accounting, water use and labor force data, allows us to fully parameterize the production technologies for each sector. Second, they allow for calculating the rate of exogenous technical change for Kazakhstan – an important parameter in economic growth models.

Hydrological data needs are annual water use by cotton and rice producers in cubic meters, and the amount of water, again in cubic meters, that flows through South Kazakhstan and through Kyzylorda. By water use, we mean the amount of water each region takes from the Syr Darya. We define the difference between the amount of water that flows down the Syr Darya and the amount withdrawn from the river equals the amount of water in cubic meters that empties into the Aral Sea.

The consumption shares were also derived directly from the SAM. See Roe, Smith and Saraçoğlu (2010) for details on this process. The consumption shares are presented in Table 4.6.

Simulations

This section presents the results of two simulations. The baseline simulation examines the economics of the status quo policy, where Kyzylorda and South Kazakhstan receive a fixed amount of water each period: rice and cotton producers receive 634 km^3 and 2,321 km^3 of Syr Darya water each year, while South Kazakhstan other agriculture (SKOA) and Kyzylorda other agriculture (KOA) each receive 1,202 km^3 and 536 km^3. The rest of Kazakhstan agriculture (ROKA) is endowed with 9,306 km^3 of water. In the baseline model we assume 60% of the basin water eventually reaches the fields. The objective of the first simulation is to establish a baseline set of results, and to understand some of the basic forces operating in the economy and how they link with land and water values. The second simulation keeps the total water allocation the same, but allows the two regions to trade water across regions. The discussion of results traces out the impact of capital deepening (i.e., the impact of an increasing capital-to-labor ratio) on agricultural and non-agricultural production over time, and on the shadow value of water. Capital deepening occurs when an economy's capital stock grows faster than its labor force.

The baseline scenario

Roe, Smith and Saraçoğlu (2010) discuss the link between factor intensity, capital deepening and economic structure and suggest capital deepening tends to favor the more capital intensive sectors. One of the effects of capital deepening is a downward pressure on rates of return to capital and upward pressure on wages, as labor becomes relatively more scarce than capital over time. Although not shown here, capital deepening is predicted to occur, and Table 4.7 summarizes the predictions

Table 4.7 Sector value-added (in US$1,000)

Year	SKOA	KOA	ROKA	Cotton	Rice	Industry	Services	GDP
2007	532,250	99,239	5,979,619	62,847	30,450	46,086,963	40,644,177	93,435,545
2012	533,488	99,470	5,993,530	63,002	30,976	58,851,302	51,317,207	116,888,975
2017	557,859	104,014	6,267,330	65,886	32,730	72,584,647	63,070,983	142,683,449
2022	600,726	112,007	6,748,922	70,954	35,505	87,641,817	76,105,877	171,315,808
2027	660,286	123,112	7,418,056	77,992	39,229	104,381,792	90,674,992	203,375,459
2032	736,286	137,282	8,271,894	86,972	43,906	123,185,702	107,077,598	239,539,640
2037	829,444	154,652	9,318,480	97,979	49,593	144,469,977	125,658,307	280,578,432
2042	941,171	175,483	10,573,698	111,178	56,379	168,697,651	146,810,904	327,366,464
2047	1,073,467	200,150	12,059,991	126,808	64,391	196,389,574	170,984,052	380,898,433
2052	1,228,873	229,126	13,805,919	145,167	73,785	228,136,268	198,689,912	442,309,050
2057	1,410,483	262,988	15,846,236	166,622	84,748	264,611,121	230,514,428	512,896,626

on sector value-added (often referred to as sector GDP). Production in each sector increases over time, with each of the more labor intensive agricultural sectors' value-added doubling in about 40 years. Value-added in the relatively capital intensive manufacturing and service sectors, however, doubles in less than 20 years, and more than quadruples in 40 years. This outcome is not too surprising, as capital deepening tends to favor the relatively more capital intensive sectors. In this case, agriculture's relatively high dependence on labor puts it at a disadvantage when competing with manufacturing and services for resources over time. Adding to agriculture's problem is the fact that land and water are fixed factors, which puts an additional drag on the agricultural sectors' ability to compete with the rest of the economy for resources. Table 4A.1 in Appendix 4.2 shows that although agricultural output increases, its importance in the economy falls, as evidenced by the decrease in agriculture's share of total value-added (GDP) over time.

Being a non-traded good, demand forces put upward pressure on service good prices, allowing them to compete better for resources, in spite of their high labor share. Industry, the most capital intensive sector in the economy, takes advantage of this position and realizes the largest rate of increase in output over the 50-year period. Unlike agriculture, both the manufacturing's and service sector's shares of GDP increase over time. Again, one of the main drivers of this structural change is the relative importance of capital across the sectors: in general, the more (less) capital intensive in a sector, the more (less) able it will be in competing for productive resources, and hence, the larger (smaller) will be its share of GDP as the economy grows.

Given that value-added for each agricultural sector increases over time, if land and water endowments remain relatively constant, it is reasonably simple to show land and water contributions to GDP must increase as the economy grows. This underscores one important link between natural resource values and growth: natural resource values are influenced by the ability of the sector using the resource to compete for capital and labor. Highly competitive sectors will attract capital and labor at a faster rate than other sectors, and in general, increase the productivity of water and land at a faster rate than less competitive sectors.

Table 4.8 examines, more closely, water's contribution to GDP. The table presents two types of water values: the *unit shadow rental rate* of water for each of the agricultural sectors, and the aggregate *shadow value added* of water for each sector. Here, the unit shadow rental rate is the amount a farmer would be willing to pay for the right to purchase an additional unit of water in the current period. This is to be contrasted with a sector's unit shadow price of water, which is the amount a farmer would pay for permanent user rights to a unit of water. The reader can verify that the rate of growth in unit shadow rental rates follows closely, the rate of growth in output in each sector.

Table 4.8 reveals that at each point in time, the unit shadow rental rates vary across sectors, with SKOA assigning a higher unit shadow rental rate of water than any of the agricultural sectors along the Syr Darya. On average, SKOA values water two and one half times more than cotton producers, and eight times more than rice producers. The estimated shadow rental rates are much smaller than those

Table 4.8 Water flow shadow values – per unit and total

Year	Unit shadow rental rate per sector ($/km³)					Shadow value-added per sector (US$ 1,000)						
	SKOA	KOA	ROKA	Cotton	Rice	SKOA	KOA	ROKA	Cotton	Rice	Total
2007	39.04	16.32	56.68	15.14	4.87	35,212	6,565	395,596	7,198	8,469	453,041
2012	39.13	16.36	56.81	15.18	4.95	35,294	6,581	396,516	7,216	8,616	454,222
2017	40.92	17.11	59.40	15.87	5.23	36,906	6,881	414,630	7,546	9,104	475,067
2022	44.07	18.42	63.97	17.10	5.67	39,742	7,410	446,491	8,127	9,875	511,645
2027	48.44	20.25	70.31	18.79	6.27	43,683	8,145	490,759	8,933	10,911	562,430
2032	54.01	22.58	78.41	20.96	7.02	48,711	9,082	547,246	9,961	12,212	627,213
2037	60.84	25.44	88.33	23.61	7.92	54,874	10,231	616,486	11,222	13,794	706,607
2042	69.04	28.87	100.22	26.79	9.01	62,265	11,610	699,528	12,734	15,682	801,818
2047	78.74	32.92	114.31	30.55	10.29	71,018	13,241	797,857	14,524	17,910	914,550
2052	90.14	37.69	130.86	34.98	11.79	81,299	15,158	913,363	16,627	20,523	1,046,969
2057	103.47	43.26	150.20	40.15	13.54	93,314	17,399	1,048,345	19,084	23,572	1,201,713

from the 2003 IWMI study. This result can be due to several reasons, but absent more details on exactly how the IWMI figures were calculated, we are unable to uncover a plausible explanation for the differences.

One policy implication of the different unit shadow rental rates across sectors is the river basin could likely benefit from reallocating water across the agricultural uses in the region. In this case, the likely outcome of such an institutional change is cotton, rice and KOA would rent some of its water to SKOA. The next section examines a possible outcome of such trades.

Another policy implication imbedded in Table 4.8 relates to the fact that the shadow rental rates will likely change over time: in this case, increase as the economy evolves. Earlier, we note this occurs because real output in each agricultural sector increases over time. More specifically, this is the result of improvements in technical change, and the corresponding increase in labor productivity. The policy implication here is, if negotiations ever emerge for trading water use rights, the mechanism that implements the water trading scheme will allow for renegotiating prices over time.

Table 4.8 also presents water's contribution to GDP – or the *flow* shadow value-added – of each sector. Here, the Syr Darya basin accounts for about 13% of water's contribution to agricultural GDP (e.g., in 2007 we have 57.45/453 = 0.1268) with SKOA being the major contributor to water's value added along the basin. These patterns hold across each point in time, and are primarily the result of the size of SKOA's initial water endowment, and its relatively high shadow rental rates. Table 4A.2, in Appendix 4.2 shows land and water's contribution to GDP is equal to about 2% in 2007, but falls to a little under 1% 50 years later: a consequence of the land and water constant being fixed, whereas the capital and labor stocks grow over the 50-year period.

As noted previously, the unit shadow values of water in Table 4.8 are measures of how much a farmer would pay to purchase an additional unit of water in given period of time. The prices and value added levels in Table 4.8 are crucial ingredients in calculating the investment (or stock) value of water: the amount a farmer would pay for permanent rights to use the water. A standard definition of the stock value of an asset is the discounted present value of all future net income streams. In the case of water (and land), the flow shadow value added at a given time is the net income stream for that period. The appropriate discount rate is given by equation (6) in Appendix 4.1, and depends on the rate of return to capital and changes in the stock of water (if any).

Inclusive wealth is a concept of human well-being gaining popularity among economists, and more recently, policy makers. Simply put, inclusive wealth is the total asset value of four types of capital: natural capital (e.g., minerals, water and land), physical/man-made capital (e.g., machinery and buildings), human capital (embodied in education) and institutional capital (e.g., patent systems and legal systems). We now examine the asset (or stock) value of water.

Table 4.9 presents the unit stock price of water for each sector. Given the shadow rental rates increase over time, it is necessarily the case that the unit shadow price in period $t + 1$ will be larger than the unit shadow price in period t. Hence, for

Table 4.9 Unit stock shadow price of water (US$/km³)

Year	"Correct" calculations					Price/earnings ratio				
	SKOA	KOA	ROKA	Cotton	Rice	SKOA	KOA	ROKA	Cotton	Rice
2007	1,123.5	469.8	1,631.0	435.9	144.8	624.8	261.2	907.0	242.3	77.9
2012	1,332.9	557.3	1,934.9	517.1	172.8	695.5	290.8	1,009.6	269.8	88.0
2017	1,572.7	657.6	2,283.0	610.2	204.6	786.0	328.6	1,141.0	304.9	100.4
2022	1,846.3	772.0	2,680.2	716.3	240.8	896.1	374.7	1,300.8	347.6	115.4
2027	2,158.6	902.6	3,133.6	837.5	282.0	1,026.9	429.4	1,490.7	398.4	132.9
2032	2,515.7	1,051.9	3,652.0	976.1	329.0	1,180.4	493.5	1,713.5	458.0	153.3
2037	2,925.0	1,223.0	4,246.1	1,134.9	382.8	1,359.3	568.4	1,973.3	527.4	177.0
2042	3,394.8	1,419.4	4,928.1	1,317.2	444.5	1,567.2	655.3	2,275.1	608.1	204.5
2047	3,934.8	1,645.2	5,712.1	1,526.8	515.4	1,808.2	756.0	2,624.9	701.6	236.2
2052	4,556.4	1,905.1	6,614.4	1,768.0	596.9	2,087.2	872.7	3,030.0	809.9	273.0
2057	5,272.4	2,204.5	7,653.9	2,045.8	690.9	2,410.0	1,007.7	3,498.6	935.1	315.4

Kazakhstan, the rate of exogenous technical change and labor force growth leads to an increase in the shadow stock price of water for each sector. Understand, however, that nothing guarantees this outcome universally: if a sector is sufficiently labor intensive, and the rate of technical change and labor force growth is relatively small, it is possible for the shadow stock price of a sector to fall over time.

Table 4.9 also presents the price/earnings ratio for each sector over time. We do this to illustrate how important it is to use the "correct" discount factor. The "correct" unit price is calculated using the discount factor in equation (6) of Appendix 4.1, while the "traditional" unit price is calculated using the standard calculation – i.e., dividing the water rental rate by the interest rate (e.g., the price/earnings, or PE, ratio). For each period in time, the PE ratio is smaller than the corresponding shadow price corrected discounted. The point to stress to technicians is, deriving the "correct" unit stock value of a natural resource *requires* exploiting a no-arbitrage condition derived from macroeconomic conditions and variables (Equation 6 in Appendix 4.1). It follows that attempting to calculate the stock value of a natural asset in a (static or dynamic) partial equilibrium setting can lead to biased unit stock price estimates. In such a case, using the PE ratio can seriously bias downward (in some cases, upward), a natural resource's contribution to national wealth.

Why would a policy maker care about the asset value of water, or land? One reason is a calculation of the stock values under the status quo policy gives her a good idea of the value of the asset under that regime. If she is considering another policy regime, almost certainly the policy change will trigger a change in the stock value. In the case of agriculture, if the stock value of land and water increases, the policy improves the income and wealth of the asset owners. If not, the policy worsens their wealth position. Given that stock values are the discounted value of current and future land and water rental payments/income, an increase (decrease) in the stock value is likely accompanied by an increase (decrease) in short run and long run farmer income. Hence, a change in that single index of value signals corresponding changes to income streams in the near and long run. One problem with this index, however, is it can hide the timing and magnitudes of the changes, suggesting if a large change in stock values is predicted, a close look at the predicted income streams is warranted.

The total asset value of land and water is given in Table 4.10. The water asset values are derived by simply multiplying the stock price of water for a sector by the quantity of Syr Darya water it is allocated. Land asset values are derived by solving equation (5) in Appendix 4.1. One thing that might not be obvious, however, is while the model pays close attention to water quantity levels and unit prices, the quantity of arable land is normalized to unity. This means we interpret land rent as the total value of rental payments to landowners, not rent per hectare.

Calculating the full inclusive wealth of Kazakhstan is beyond the scope of this analysis. One can, however, compare the wealth derived from water and land with the wealth derived from physical capital, as the no arbitrage condition implicitly defines the asset value of land and water in terms of the value of physical capital – whose unit price is normalized to unity. Table 4A.3 in Appendix 4.2 shows the asset value of land and water is 17% that of physical capital in 2007, and drops down to 12% in 2057. These values warrant a closer look, and considering we have not included the asset values of minerals and crude oil, suggest natural resources hold a prominent position in Kazakhstani wealth.

One reason for this study is to develop a model for measuring the value of natural resources, in this case water, in an economy. The other objective is to discuss ways to "mainstream" the economic information given previously into the policymaking process. We conclude this section with a summary of what the information in Tables 4.6 through 4.10 conveys to the astute policy maker. First, if predicted unit shadow water rents increase over time, the agricultural ministry can be confident that agriculture is reasonably competitive with manufacturing and services in the markets capital and labor (factor markets) – competitive enough to not contract as the economy grows. Second, if the unit shadow water rents vary across sectors, it may be worthwhile setting up a commission to investigate the potential gains from water trading. Third, the stock value of water and land is a scalar index of the value of current and future income farmers will receive from land rent and shadow water rent. These values, almost certainly, are influenced by the agricultural policy environment: industrial policy can affect these stock values, too – especially if an industrial policy gives manufacturing or services an edge in competing for capital and labor. In any event, the careful minister will insist on conducting an exercise that estimates the stock values of the assets under her purview, given the proposed policy change. If the stock values of land and water increase with a proposed policy (e.g., water trading or purchasing Syr Darya water for Aral Sea restoration), one can cautiously assume farmers, and possibly society in general, are benefitting from the policy. An additional look at the aggregate rental values over time can increase her level of confidence in supporting the policy; if for each year, the water and land rental values under the new policy regime are higher than those

Table 4.10 The stock value of water and land

Year	Stock value of water (US$/km³)					Stock value of land (US$/km²)				
	SKOA	KOA	ROKA	Cotton	Rice	SKOA	KOA	ROKA	Cotton	Rice
2007	810,604	151,139	9,106,853	165,756	201,678	2,600,934	484,950	29,220,489	238,626	36,273
2012	961,658	179,304	10,803,893	196,655	240,633	3,085,612	575,319	34,665,656	283,108	43,279
2017	1,134,667	211,562	12,747,600	232,042	284,954	3,640,738	678,823	40,902,285	334,053	51,251
2022	1,332,080	248,370	14,965,463	272,419	335,325	4,274,164	796,927	48,018,580	392,181	60,310
2027	1,557,426	290,386	17,497,143	318,509	392,675	4,997,216	931,742	56,141,793	458,532	70,625
2032	1,815,113	338,433	20,392,170	371,212	458,144	5,824,041	1,085,905	65,430,853	534,405	82,400
2037	2,110,378	393,486	23,709,374	431,600	533,073	6,771,440	1,262,550	76,074,521	621,341	95,877
2042	2,449,318	456,682	27,517,245	500,920	619,015	7,858,975	1,465,323	88,292,556	721,136	111,334
2047	2,838,972	529,334	31,894,880	580,612	717,760	9,109,236	1,698,437	102,338,751	835,862	129,094
2052	3,287,448	612,954	36,933,352	672,334	831,366	10,548,233	1,966,741	118,505,326	967,907	149,526
2057	3,804,071	709,280	42,737,430	777,993	962,196	12,205,888	2,275,814	137,128,444	1,120,015	173,057

under the old regime, she can safely conclude the policy is beneficial to farmers. If predicted rental values are higher in some years, but not in others, then a more careful consideration of the tradeoffs is warranted.

Oblast water trading

One of the main points to take away from the previous section is: policy can influence both the flow and the asset value of natural resources. This section takes a closer look at the potential benefits of allowing oblasts to trade water use rights among themselves, over time. As in the prior section, the results assume the Syr Darya basin has been allocated about 6,980 km³ of water, and on average, 60% of that water eventually reaches the fields. The overall economy dynamics here are similar to those in the base scenario, and aside from a quick comparison of sector value added levels, are not discussed in this section. Instead, we focus attention on the unit and total, flow and stock prices of water across the agricultural sectors along the Syr Darya. One thing we will look for is whether a water trading environment increases, or decreases, the asset value of water.

Table 4.8 suggests, on average, SKOA producers place the highest value on water, and if presented a chance to purchase – either per period or permanently – water rights, they would seize the opportunity. If water trading occurred, the expected outcome would be for a single price to emerge such that the last unit of water used in each sector would have the same value – economists would write that the marginal value product of water is the same for each sector. The rationale underlying this outcome goes something like this: if SKOA was willing to pay a little more than everyone else for an additional unit of water, then someone along the river basin (say one of the rice producing oblasts) would be willing to give up some of their water to SKOA. Doing so would make water a little scarcer in the rice region, and hence, push up the (marginal) value of water. It would make water a little less scarce in SKOA, and put downward pressure on its (marginal) value of water. Barring one or more sectors facing irrigation system constraints in delivering additional water, this process would tend to nudge water trading prices along the river basin towards a single trading value.

Perhaps as a reminder, sector value-added is measured as the sum of payments to labor, capital, land and water used to produce output in each sector. Table 4.11 shows that, as in Table 4.6, as the economy grows, value-added in each sector increases over time. Simple calculations will show, for each sector, the rates of growth between 2007 and 2057 are similar, with all but rice and services growing slightly faster with water trading. The rice sector virtually closes, but most of the foregone production income is recovered from income it gets from renting water to SKOA.

Table 4A.4 in Appendix 4.2 reveals Kyzylorda produces very little rice, and Table 4.12 reveals it earns almost all of its value-added on water rent received from SKOA oblasts. Table 4.11 suggests that each year, SKOA producers would purchase water from each of the other agricultural sectors along the Syr Darya, buying almost all of the rice water available each year. For example, in 2007,

Table 4.11 Sector value-added with water trading (in US$1,000)

Year	SKOA	KOA	ROKA	Cotton	Rice	Industry	Services
2007	690,678	98,134	5,963,110	58,376	31	45,948,101	40,734,542
2012	692,846	98,442	5,981,865	58,572	35	58,687,630	51,403,872
2017	724,901	102,996	6,258,656	61,292	39	72,390,633	63,154,524
2022	780,908	110,954	6,742,235	66,034	44	87,412,238	76,187,424
2027	858,568	121,988	7,412,769	72,607	50	104,111,201	90,755,992
2032	957,579	136,056	8,267,637	80,985	57	122,868,084	107,159,681
2037	1,078,887	153,291	9,315,010	91,247	65	144,098,495	125,743,232
2042	1,224,339	173,958	10,570,844	103,552	75	168,264,438	146,900,540
2047	1,396,540	198,425	12,057,630	118,119	86	195,885,524	171,080,388
2052	1,598,802	227,163	13,803,956	135,228	99	227,550,824	198,795,082
2057	1,835,153	260,744	15,844,601	155,220	114	263,932,039	230,630,744

Table 4.12 Water trading levels (supply) and values by sector

Year	Unit rental price/km³	Levels (km³)				Values (1000 US $)			
		SKOA	KOA	Cotton	Rice	SKOA	KOA	Cotton	Rice
2007	20.91	−1,464.0	11.2	60.5	1.392.3	−30,608	235	1,265	29,110
2012	20.97	−1,463.9	11.2	60.4	1,392.2	−30,704	236	1,268	29,201
2017	21.95	−1,463.8	11.2	60.4	1,392.2	−32,124	247	1,325	30,552
2022	23.64	−1,463.8	11.2	60.4	1,392.2	−34,606	266	1,427	32,913
2027	25.99	−1,463.8	11.3	60.4	1,392.2	−38,047	292	1,569	36,186
2032	28.99	−1,463.7	11.3	60.3	1,392.2	−42,434	326	1,749	40,359
2037	32.66	−1,463.7	11.3	60.3	1,392.1	−47,810	368	1,970	45,472
2042	37.07	−1,463.7	11.3	60.3	1,392.1	−54,255	417	2,236	51,602
2047	42.28	−1,463.7	11.3	60.3	1,392.1	−61,886	476	2,550	58,860
2052	48.40	−1,463.7	11.3	60.3	1,392.1	−70,849	545	2,919	67,385
2057	55.56	−1,463.7	11.3	60.3	1,392.1	−81,322	626	3,351	77,346

Note: Positive values represent supply, negative values represent demand.

SKOA pays the other sectors over US$30 million for 1,464 km³ of water, with more than US$29 million used to purchase water from the rice producing oblasts.

These results, however, should not be taken too seriously, as the magnitude of water trades between Other-Ag SK and rice producers is unlikely, and is the result of the functional form (Cobb-Douglas) used in the numerical model: the specification yields an outcome where a small change in water rental rates leads to large swings in water demand. Gemma and Smith (2019) examine the effect of different production and utility function specifications on rice production quota trading in Japan.[1] We suspect a re-specification of the production technologies along the lines

Table 4.13 Sector value-added gains and losses with water trading (in US$1,000)

Year	SKOA	KOA	Cotton	Rice	Total income lost	Surplus to SKOA
2007	158,652	−839	−3,186	−1,300	−5,325	153,327
2012	159,585	−761	−3,142	−1,731	−5,634	153,951
2017	167,254	−741	−3,250	−2,130	−6,122	161,132
2022	180,374	−760	−3,476	−2,541	−6,776	173,597
2027	198,454	−807	−3,801	−2,986	−7,595	190,860
2032	221,447	−878	−4,225	−3,484	−8,587	212,860
2037	249,580	−973	−4,749	−4,051	−9,773	239,806
2042	283,289	−1,091	−5,380	−4,697	−11,168	272,122
2047	323,181	−1,234	−6,130	−5,441	−12,805	310,376
2052	370,024	−1,405	−7,012	−6,297	−14,714	355,310
2057	424,754	−1,607	−8,044	−7,284	−16,935	407,819

of Gemma and Smith (2019) will yield similar market water rental rates predicted in the current model, but much smaller water trading levels.

Returning to the model results, the entries in Table 4.13 give the level differences between agricultural value added in the market and base scenarios. The major observation to make here is (in 2007) water trading leaves cotton, rice and Kyzylorda other agriculture worse off by a little over US$5.3million, but SKOA gains over US$158 million. These values give any ministries involved in implementing such a program, the benefit side, of a cost-benefit analysis of the program. The SKOA gain leaves ample room for SKOA to compensate the KOA, cotton and rice oblasts for their lost income.

Introducing a lump sum tax on water trades might be one way to raise compensation income: for example, if estimates suggest cotton, rice and other agricultural production in Kyzylorda would lose $10 million if a water trading scheme was implemented, and SKOA was composed of 100 (for the present, identical) oblasts, then charge an purchase entry fee of $100,000 per oblast, and distribute the proceeds across the cotton, rice and KOA oblasts proportionally (according to their projected losses). This would still leave the SKOA oblasts with a generous program surplus. Of course, other sharing arrangements can be envisioned, and could generate quite a bit of dialogue along the river basin.

One outcome of water trading is, at each point in time, farmers across the Syr Darya pay the same unit rental price of water. Again, since the rental rates increase over time, it follows that the unit stock price of water will increase over time, as will the total land rent for each sector. Table 4.14 presents the trajectory of unit stock water prices, and the corresponding PE ratios. As with the base model, the "correct" unit stock price is a little more than twice that of the corresponding PE ratio. Again, this has implications for inclusive wealth analysis: PE based wealth values will likely underestimate water's contribution to the economy.

Earlier it was suggested that policy could, in principle, affect asset values. Table 4A.5 in Appendix 4.2 presents the stock value of land and water across sectors

Table 4.14 Unit stock price of water (US$/km³)

Year	"Correct"	PE ratio
2007	482.9	268.2
2012	572.9	298.6
2017	675.9	337.6
2022	793.4	384.9
2027	927.6	441.1
2032	1,081.0	507.1
2037	1,256.8	584.0
2042	1,458.6	673.3
2047	1,690.6	776.8
2052	1,957.7	896.7
2057	2,265.3	1,035.4

under the water trading scenario. Table 4.15 summarizes the potential sector gains and losses in water and land wealth when implementing a water market policy. Overall, a more "efficient" allocation of water increases the wealth value of the natural assets – or worded differently – increases the wealth value of the ecosystem services provided by the assets. Table 4.15 also reveals policy can have distributional impacts. In each of the sectors along the Syr Darya, a trade-off occurs between water and land wealth. In the case considered here, water trading establishes an equilibrium water rent rate that is lower than the SKOA's shadow rental rate in the base scenario, but higher than the shadow rental rates that prevail in the cotton, rice and KOA sectors. This makes SKOA's water endowment less valuable, but increases the value of water in the other sectors. With cheaper water, SKOA demands more water, and more water enables SKOA to compete more effectively in the capital and labor markets. These forces all contribute to increasing land productivity, and hence, the asset value of SKOA land increases.

On the other hand, shadow unit water prices increase and the total asset value of water increases for the cotton, rice and KOA sectors. Trading away some of their water to SKOA, however, puts downward pressure on the productivity of capital, labor and land. Hence, land rental rates and land wealth falls. In each sector, the gains dominate the losses, and overall welfare improves. If land and water use rights are typically held by the same individual(s), then "one hand washes the other" and the distributional impact of the policy is neutral. If resource use rights are held by different parties, as in Texas and other parts of the United States, then the policy could have unpopular consequences. Hence, if income distribution is an important policy consideration, even if a policy improves wealth indices, ministries may want to look closely at the distributional impact of the policy: if implementing a policy increases asset wealth, almost certainly enough income will be generated to compensate the loser. Table 4A.6 in Appendix 4.2 summarizes the net gains in water and land wealth given water trading.[2]

Table 4.15 Gains and losses in asset value with water trading (US$1,000)

Year	Stock value of water					Stock value of land					Total Gain
	SKOA	KOA	ROKA	Cotton	Rice	SKOA	KOA	ROKA	Cotton	Rice	
2007	−462,186	4,232	4,416	17,889	470,870	674,651	−24,446	58,914	−34,397	−42,428	667,514
2012	−548,343	5,007	4,340	21,197	557,184	774,474	−33,815	80,327	−45,057	−52,042	763,272
2017	−647,034	5,891	4,005	24,981	656,322	887,802	−44,728	99,914	−57,418	−63,069	866,665
2022	−759,647	6,897	3,573	29,301	769,639	1,015,854	−57,415	116,501	−71,710	−75,693	977,299
2027	−888,193	8,047	3,124	34,233	899,140	1,160,884	−72,111	129,377	−88,193	−90,145	1,096,164
2032	−1,035,185	9,364	2,698	39,875	1,047,345	1,325,992	−89,055	137,899	−107,140	−106,692	1,225,100
2037	−1,203,609	10,873	2,310	46,343	1,217,257	1,515,140	−108,479	141,348	−128,831	−125,634	1,366,716
2042	−1,396,943	12,608	1,966	53,769	1,412,378	1,733,294	−130,585	138,984	−153,533	−147,297	1,524,641
2047	−1,619,200	14,604	1,666	62,309	1,636,758	1,986,684	−155,517	130,297	−181,472	−172,024	1,704,105
2052	−1,875,006	16,902	1,408	72,141	1,895,062	2,283,189	−183,308	115,621	−212,793	−200,168	1,913,047
2057	−2,169,680	19,551	1,186	83,468	2,192,659	2,632,921	−213,799	97,438	−247,501	−232,077	2,164,168

The ratio of land and water wealth to normalized capital stock wealth is very close to that found in the base scenario – starting at 18% of physical asset wealth and falling to 13% (see Table 4A.7 in Appendix 4.2).

We summarize this section by reminding the reader that the analysis eventually focused on the wealth value of land and water, and used the indices to understand the impact of a natural resource policy on natural asset wealth. The primary purpose of the simulation was to illustrate how natural asset (or ecosystem service) valuation can be used to guide and understand the impact of policy. Although the illustration was natural resource policy, the notion that policy can affect natural resource wealth or the value of ecosystem service flows extends to almost any economic policy, be it trade, fiscal or industrial policy in general.

Improving irrigation efficiency

This section reports the results of a simulation where irrigation efficiency improves for each of the agricultural sectors along the Syr Darya. We begin with the baseline efficiency assumption of 60% of the IWD reaching crops, and gradually increase irrigation efficiency over a 50-year period. By 2057, 85% of the IWD reaches the crops. Aside from Syr Darya agriculture, the sector value added levels and shares follow closely those of the base model (see Tables 4A.8 and A.9 in Appendix 4.2). As in the base and water market models, production in each sector increases over time, agricultural value-added doubling in about 40 years, value-added in manufacturing and service sectors double in less than 20 years.

Table 4.16 summarizes differences between the base model results and the model with improved irrigation efficiency. Improved irrigation efficiency increases the effective water endowment of each sector, which increases the productivity of capital, labor and land in Syr Darya agricultural production: in turn, the region's

Table 4.16 Percentage difference in sector value-added: $\left(\dfrac{\text{efficiency improvement}}{\text{base}} - 1\right)$

Year	SKOA	KOA	ROKA	Cotton	Rice	Industry	Services	GDP
2007	−0.002	−0.002	−0.002	−0.002	−0.002	0.009	−0.004	0.003
2012	0.010	0.010	0.001	0.018	0.036	0.005	−0.004	0.001
2017	0.021	0.021	0.002	0.036	0.072	0.003	−0.004	0.000
2022	0.032	0.032	0.003	0.053	0.108	0.001	−0.004	−0.001
2027	0.041	0.041	0.004	0.069	0.143	0.000	−0.004	−0.002
2032	0.050	0.050	0.004	0.085	0.178	−0.001	−0.004	−0.002
2037	0.059	0.059	0.004	0.100	0.213	−0.002	−0.003	−0.002
2042	0.067	0.067	0.004	0.115	0.247	−0.003	−0.003	−0.002
2047	0.075	0.075	0.004	0.129	0.280	−0.003	−0.002	−0.002
2052	0.082	0.082	0.003	0.143	0.314	−0.004	−0.002	−0.002
2057	0.089	0.089	0.003	0.156	0.347	−0.003	−0.001	−0.002

Table 4.17 Total rate of growth in sector value-added between 2007 and 2057

Growth	SKOA	KOA	ROKA	Cotton	Rice	Industry	Services	GDP
Base	1.650	1.650	1.650	1.651	1.783	4.742	4.672	4.489
Improved irrigation efficiency	1.892	1.892	1.662	2.072	2.755	4.674	4.684	4.466

ability to compete for capital and labor increases over time. These forces enable Syr Darya agriculture to increase its production relative to base model levels.

A by-product of this enhanced ability to compete for resources is revealed in Table 4.17, which shows the total rate of growth in sector value-added over the 50-year period, 2007–2057. The rate of growth in Syr Darya agricultural output increased at the expense of industry and service sector growth (slightly). This result reveals an important aspect of ecosystem service and natural resource valuation: in general, resource abundance should enhance the competitiveness of sectors drawing on its services.

Table 4.18 compares unit shadow rent levels and total shadow rental value levels across sectors (see Table 4A.10 in Appendix 4.2 for the irrigation efficiency model's unit and total shadow rental values). With water increasingly more abundant relative to the base model, water unit shadow rental rates are lower at each point in time (imperceptibly in the initial period), across each sector. For each sector, the total shadow rental value – or shadow value-added – is higher in the increased irrigation efficiency model relative to the base case: this occurs because the rate of growth in water quantities is larger than the rate of decline in the unit shadow rental rate. Increasing the sector water endowments leads to a wider variation in shadow rental rates, suggesting again that trading water use rights might lead to improved farmer and aggregate welfare.

Table 4.19 presents the percentage difference in land values and unit water prices for each sector (see Tables 4A.11 and 4A.12 in Appendix 4.2 for the irrigation model's land and unit water stock prices). Again, since water is more abundant, it is less scarce, and the unit stock price of water is initially lower with more efficient irrigation. On the other hand, more water makes land more productive, and hence, increases land values. Given the structure of agricultural production, improvements in irrigation technology lead to a nominal increase in natural asset wealth.

Table 4.20 gives the difference in water and land stock values. With increased irrigation efficiency, farmers realize a net gain in wealth relative to the base scenario. These results suggest investing in irrigation infrastructure repair should yield benefits to Syr Darya agriculture. These values, however, likely underestimate the gain from improving irrigation efficiency. As modelled here, increased water endowments are spread over the same level of cultivated area. A more realistic specification would allow cultivated area to increase with improved irrigation efficiency. In addition to increasing the number of hectares earning rent, this modelling adjustment would have increased the productivity of water: doing

Table 4.18 Percentage difference in water shadow rental values – per unit and total: $\left(\dfrac{\text{efficiency improvement}}{\text{base}} - 1\right)$

Year	Unit shadow rental rate per sector ($/km³)					Shadow value-added per sector (US$1,000)					Total
	SKOA	KOA	ROKA	Cotton	Rice	SKOA	KOA	ROKA	Cotton	Rice	
2007	−0.002	−0.002	−0.002	−0.002	−0.002	−0.002	−0.002	−0.002	−0.002	−0.002	−0.002
2012	−0.030	−0.030	0.001	−0.023	−0.006	0.010	0.010	0.001	0.018	0.036	0.002
2017	−0.057	−0.057	0.002	−0.044	−0.010	0.021	0.021	0.002	0.036	0.072	0.006
2022	−0.083	−0.083	0.003	−0.064	−0.015	0.032	0.032	0.003	0.053	0.108	0.009
2027	−0.108	−0.108	0.004	−0.084	−0.020	0.041	0.041	0.004	0.069	0.143	0.011
2032	−0.131	−0.131	0.004	−0.102	−0.025	0.050	0.050	0.004	0.085	0.178	0.013
2037	−0.153	−0.153	0.004	−0.120	−0.030	0.059	0.059	0.004	0.100	0.213	0.015
2042	−0.174	−0.174	0.004	−0.137	−0.035	0.067	0.067	0.004	0.115	0.247	0.016
2047	−0.194	−0.194	0.004	−0.153	−0.040	0.075	0.075	0.004	0.129	0.280	0.018
2052	−0.213	−0.213	0.003	−0.169	−0.045	0.082	0.082	0.003	0.143	0.314	0.019
2057	−0.231	−0.231	0.003	−0.184	−0.049	0.089	0.089	0.003	0.156	0.347	0.020

Table 4.19 Percentage difference in stock land value and unit stock shadow price of water: $\left(\dfrac{\text{efficiency improvement}}{\text{base}} - 1\right)$

Year	Unit stock price of water					Stock price of land					Total
	SKOA	KOA	ROKA	Cotton	Rice	SKOA	KOA	ROKA	Cotton	Rice	
2007	-0.1253	-0.1253	-0.0015	-0.0995	-0.0288	0.0425	0.0425	-0.0015	0.0762	0.1709	-0.0011
2012	-0.1143	-0.1143	-0.0019	-0.0822	0.0058	0.0508	0.0508	-0.0019	0.0912	0.2037	-0.0002
2017	-0.1002	-0.1002	-0.0021	-0.0625	0.0412	0.0578	0.0578	-0.0021	0.1038	0.2319	0.0007
2022	-0.0837	-0.0837	-0.0022	-0.0411	0.0772	0.0639	0.0639	-0.0022	0.1147	0.2564	0.0018
2027	-0.0653	-0.0653	-0.0021	-0.0180	0.1136	0.0691	0.0691	-0.0021	0.1242	0.2777	0.0030
2032	-0.0451	-0.0451	-0.0020	0.0064	0.1505	0.0737	0.0737	-0.0020	0.1323	0.2961	0.0042
2037	-0.0232	-0.0232	-0.0018	0.0323	0.1878	0.0776	0.0776	-0.0018	0.1391	0.3117	0.0055
2042	0.0007	0.0007	-0.0015	0.0597	0.2257	0.0808	0.0808	-0.0015	0.1446	0.3244	0.0069
2047	0.0266	0.0266	-0.0012	0.0888	0.2640	0.0832	0.0832	-0.0012	0.1488	0.3339	0.0082
2052	0.0548	0.0548	-0.0009	0.1197	0.3029	0.0848	0.0848	-0.0009	0.1516	0.3400	0.0095
2057	0.0855	0.0855	-0.0007	0.1527	0.3424	0.0855	0.0855	-0.0007	0.1527	0.3424	0.0108

Table 4.20 Water and land wealth in the water market and base scenarios (US$1,000)

Year	Irrigation Efficiency Scenario			Base Scenario			Difference (efficiency – base)		
	Water	Land	Total	Water	Land	Total	Water	Land	Total
2007	10,279,294	32,691,971	42,971,265	10,436,037	32,581,271	43,017,308	-156,743	110,700	-46,043
2012	12,216,335	38,806,993	51,023,328	12,382,151	38,652,974	51,035,124	-165,815	154,019	-11,796
2017	14,446,075	45,816,223	60,262,297	14,610,835	45,607,150	60,217,985	-164,761	209,073	44,312
2022	17,003,280	53,821,401	70,824,681	17,153,669	53,542,162	70,695,832	-150,389	279,239	128,850
2027	19,937,100	62,967,051	82,904,151	20,056,152	62,599,908	82,656,060	-119,052	367,143	248,091
2032	23,308,956	73,432,802	96,741,758	23,375,087	72,957,604	96,332,692	-66,131	475,198	409,066
2037	27,192,303	85,431,449	112,623,752	27,177,929	84,825,729	112,003,659	14,374	605,720	620,093
2042	31,673,472	99,210,083	130,883,555	31,543,201	98,449,324	129,992,526	130,270	760,759	891,029
2047	36,853,207	115,052,992	151,906,199	36,561,583	114,111,380	150,672,963	291,623	941,612	1,233,236
2052	42,848,716	133,285,584	176,134,299	42,337,482	132,137,733	174,475,214	511,234	1,147,851	1,659,085
2057	49,796,098	154,278,629	204,074,727	48,991,002	152,903,219	201,894,220	805,097	1,375,410	2,180,507

so would, without question, increase the stock value of the ecosystem (natural resources). Viewed another way, Table 4.18 also provides a measure of another policy: increasing irrigation efficiency and then sending the additional water (reaching plants) down the Syr Darya to the Aral Sea. In this case, if saved water was sent down to the North Aral Sea, agriculture would forgo an additional $14.5 million in natural asset wealth.

Finally, the ratio of land and water wealth to normalized capital stock wealth is similar to that estimated in the base and market scenarios – starting at around 17% of physical asset wealth and falling to a little over 12% (see Table 4A.13 in Appendix 4.2).

We conclude this section by noting the stock value of natural resources seems promising as an economic index of potential value to policy makers, and exploits the fact that, in general, policy choices affect the underlying wealth embedded in ecosystems and natural resources. Our understanding is this is a relatively new use (or at least actual application) of natural asset stock values: as such, one should view our interpretations with a critical eye. Still, it appears that measuring the impact of a proposed policy on natural asset values has potential as a tool for main-streaming ecosystem services into regional and macroeconomic policy debate.

Conclusions

The major objective of this chapter has been to develop an analytical tool for main-streaming ecosystem services into Syr Darya water management. In this case, the analytical tool measures the contribution of water and land to agricultural value added over time – i.e., the flow value of the provisioning services provided by the natural resources. The analytical tool uses these flow values to calculate the wealth, or stock value of the provisioning services provided by the two resources. The ability to measure the flow and stock value of the ecosystem services and natural assets forms the core of an analytical framework for understanding the impact of policy on ecosystems and natural assets.

The analysis first asks, does a policy increase the stock value of an ecosystem? If so, then the policy is likely to be welfare increasing – a positive in terms of sustainability. It then asks if there are distributional impacts of the policy. In the case of agricultural production, if water use rights and land use rights are owned by different agents, even if a policy is welfare enhancing (improves total welfare), one group of agents can be made better off at the expense of another. If a policy is welfare enhancing, we then examine how the income of agents evolve over time with the new policy, and compare it with the income of the status quo (or another) policy. If the total flow value at each point in time under the new policy is higher than the corresponding value under the status quo, the policy has the potential to unambiguously improve total welfare (with side payments). To summarize, the previous analytical approach provides a framework for conducting a type of cost-benefit analysis in general equilibrium and economy-wide settings.

The empirical applications presented previously build on the conceptual model outlined in Smith (2014) – a model having its roots in the dynamic, general equilibrium models introduced in Roe, Smith and Saraçoğlu (2010). Perhaps one of the most important features of the tool is its use of natural asset wealth measures to understand the impact of policy. Another side benefit of the analytical tool is its departure from using the price/earnings ratio as an estimate of (natural) asset value – an approach that can underestimate asset values by at least one half its value when calculated properly. If interested in estimating inclusive or comprehensive wealth, the P/E ratios will typically lead to an underestimate of the loses (gains) from poorly (smartly) managing ecosystems.

The chapter uses the mainstreaming tool to analyze the impact of three policy scenarios: the status quo policy of allocating a fixed amount of water across four agricultural sectors along the Syr Darya; a policy that allows oblasts along the Syr Darya to trade water use rights among themselves each year; and a policy that improves the efficiency in which irrigation water is delivered to the field. Results suggest significant welfare/income gains might exist if the water authorities along the Syr Darya basin could implement an efficient water trading mechanism. Current results suggest more modest gains would be realized when improving irrigation efficiency – with a warning that, almost certainly, the current model setup is underestimating the potential gains from such a policy.

The dynamic nature of the analysis can be used to gain insights into links between the short-run economic gains from overexploiting an ecosystem today (e.g., irrigating rice and cotton along the Syr Darya) and the long-run losses caused by that overexploitation (Aral Sea demise). This type of analysis, however, would require close collaboration between economists, hydrologists and other natural resource scientists. Examples of smaller ecosystems that could go the way of the Syr Darya include Mono Lake and the Salton Sea in California. Other water ecosystems under stress include the Hai River Basin in China, the Ogallala aquifer system in the Great Plains and the Punjab aquifer system in India (see Smith, Nelson and Roe, 2015, for an analysis of the Punjab aquifer system). Past and current groundwater withdrawals are believed to eventually have deleterious effects on groundwater tables (e.g., saltwater intrusion in Punjab) and corresponding negative economic impacts on the regions relying on the resources.

The point to make here is, mainstreaming efforts should lend insight into how current resource demands might affect the level of ecosystem services available in the future. Furthermore, efforts should be made to understand the links, if any, between how resource management in one region affects resource availability in another. Related contemporary issues include water management along the Nile and the Mekong river basins. Increasing water demand by China will likely have economic impacts on the downstream countries, as well as impacts on their ecosystems. A clear understanding of the hydrological, ecosystem and economic trade-offs along these river basins could benefit from an approach similar to the one introduced in this chapter.

Although not discussed here, given total water and land stock values increase over time, one would likely conclude current water use is "sustainable."[3] This conclusion highlights the importance of developing a mainstreaming tool that actually measures what policy makers want measured. For example, if the objective is to understand whether a given policy supports sustainable natural asset values, the previous models are sufficient. If the objective is to understand if Syr Darya and Aral Sea water management is sustainable, the models are not up to the task. They would conclude agricultural production is a sustainable use of water and allow the kind of decimation of the Aral Sea observed between the 1960s and the 1990s. A mainstreaming tool with the features alluded to in the previous paragraph would have given planners and policy makers a clearer understanding of the cost of agricultural production.

Another topic for future research is that of more carefully measuring the agricultural production technologies: the more accurate is the measure of how output levels change when we use an additional unit of water, the more accurate will be the model's estimation of land and water, flow and stock values. This project requires collecting (or analysing existing) data on agricultural production: e.g., oblast or farm level data on the quantity of rice produced, hectares planted to rice, labor used in rice production, a measure of the capital used in rice production, quantity of pesticide used in rice production, quantity of fertilizer used, and most importantly, the quantity of water applied. Ideally, this data would be collected over several years, and the panel would be used to measure a simple Cobb-Douglas function for the crops of interest (see Smith and Gemma, 2014).

With minor revisions, the model can examine the impact of traditional macro-economic (trade or industrial) policy on natural asset contribution to sector and total value-added, and the corresponding stock values. The current model aggregates Kyzylorda and South Kazakhstan manufacturing and services into an "all Kazakhstan" manufacturing and "all Kazakhstan" service sector. Still, it is possible to disaggregate manufacturing (and services) into Kyzylorda manufacturing, South Kazakhstan manufacturing and the rest of Kazakhstan manufacturing, but we felt this was an unnecessary complication given the report's focus on water management policy.

Acknowledgements

The authors acknowledge the support, encouragement and the provision of data by the staff from the Regional Environmental Centre for Central Asia (CAREC), and the cooperation of professionals and experienced personnel from the Committee of Water Resources, the Kazakhstan Ministry of Agriculture. This report would not have been possible without their help and guidance. Feedback provided by attendees to the February 23–24, 2016, conference in Geneva Switzerland titled: "Mainstreaming of Ecosystem Service into Sectoral and Macroeconomic Policies and Programs in the Republic of Kazakhstan" is appreciatively acknowledged. Errors and omissions remain the responsibility of the authors.

Appendix 4.1

The economic model, baseline scenario[4]

Denote the economy's time t endowment of physical capital, labor, land and water by $K(t)$, $L(t)$, Z and $H(t)$ respectively.[5] Firms use these factors to produce five final goods: three agricultural goods (cotton, rice and other agriculture), manufactured goods and service goods. In what follows, $K_j(t)$, $L_j(t)$, $H_j(t)$ and Z_j represent the amount of an input used by sector j – e.g., K_j and L_j are the respective amounts of capital and labor demanded by sector j. As a warning, most variables that follow are functions of time, but in an effort to minimize mathematical notation, we almost always drop the time variable. For example, although the amount of capital demanded by cotton producers can vary over time, we will typically represent it by K_{a1} instead of $K_{a1}(t)$.

Capital and labor are used by each sector, and are mobile across the economy. Land is used by agricultural production only, and is fixed to a region or sector, where shortly we discuss when it is possible to view a sector and region as equivalent. Given the previous discussion, water will initially be viewed as a sector specific resource, but relaxed when examining water policy options. We assume all final good markets are perfectly competitive, with the implication being no single consumer or producer can influence market prices.

The structure of agricultural production is such that South Kazakhstan is the major region producing cotton, while Kyzylorda is the major region producing rice. Both regions produce other agriculture, and as noted previously, we aggregate all non-cotton production into a South Kazakhstan other agriculture, and all non-rice production into a Kyzylorda other agriculture. Let Y_{a1} represent cotton production in South Kazakhstan, and Y_{a2} as rice production in Kyzylorda. We then let Y_{a31} represent other agricultural production in South Kazakhstan, Y_{a32} represent other agricultural production in Kyzylorda, and Y_{a33}, all other agricultural production in the rest of Kazakhstan. The unit price of cotton and rice are denoted p_{a1} and p_{a2}, respectively, and the unit price of the rest of agriculture is denoted p_{a3}. A single price for other agriculture implies households make no distinction between the composite "other agricultural goods". We assume each agricultural price is exogenous and determined by world prices.

Denote manufactured and service output by $Y_m(t)$ and $Y_s(t)$, respectively. The manufacturing good is traded internationally at exogenous world price p_m, while the service good is non-traded and traded at the endogenous price $p_s(t)$. Our current

understanding is that manufacturing contributes a relatively small amount to Kyzylorda or South Kazakhstan gross domestic product (GDP). Also, manufacturing
uses very little water drawn from the Syr Darya. Assuming these two conditions
hold, we integrate Kyzylorda and South Kazakhstan manufacturing GDP into the
aggregate manufacturing sector for all of Kazakhstan. The same is true for the
service sector. Hence, the modelled economy has seven productive sectors, but
five final goods.

Household preferences, savings and consumption Represent utility with a
Cobb-Douglas function, namely

$$U\left(Q_{a2}, Q_{a3}, Q_m, Q_s\right) = Q_{a2}^{\gamma_{a2}} Q_{a3}^{\gamma_{a3}} Q_m^{\gamma_m} Q_s^{\gamma_s}$$

Here, $Q_{a2}(t)$, $Q_{a3}(t)$, $Q_m(t)$ and $Q_s(t)$ represent the quantity (indices) of rice, other
agriculture, manufacturing and services consumed by households. The parameters
$\gamma_j, j = a2, a3, m, s$, represent the share of household expenditure spent on good j.

The expenditure function associated with $U(\cdot)$ is defined as

$$E\left(p_{a2}, p_{a3}, p_m, p_s, \mu\right) \equiv \min_{Q_{a2}, Q_{a3}, Q_m, Q_s} \left\{ \begin{array}{l} p_{a2}Q_{a2} + p_{a3}Q_{a3} + p_m Q_m \\ + p_s Q_s : \mu \le Q_{a2}^{\gamma_{a2}} Q_{a3}^{\gamma_{a3}} Q_m^{\gamma_m} Q_s^{\gamma_s} \end{array} \right\}$$

The expenditure function derived from any strictly concave, twice (continuously)
differentiable utility function is itself twice continuously differentiable, as well as
being non-decreasing and concave in final good prices.

The household's optimal savings decision solves the following dynamic optimization problem:

$$\max_{\mu(t)} \int_0^\infty \ln \mu(t) e^{-\rho t} dt$$

subject to; (1) the initial conditions $K(0), H(0), Z, L(0)$; (2) the flow budget constraint (suppressing the time argument)

$$\dot{K} = wL + rK + \Pi^{a3}\left(p_{a3}, r, w, Z_{a3}\right) + \sum_{j=a1,a2} \Pi^{ja}\left(p_a, r^k, w, Z_j, H_j\right)$$

$$- E\left(p_{a2}, p_{a3}, p_w, p_s, u\right)$$

and (3) the transversality condition

$$\lim_{t \to \infty} \left\{ K(t) e^{-\int_0^t r^k(v)dv} \right\} = 0$$

Here $\dot{K}(t) = dK(t)/dt$ is the time derivative of the capital stock, δ is the rate of
capital depreciation and ρ is the household's discount factor.

The solution to this optimization problem yields the Euler condition:

$$\frac{\dot{\mu}(t)}{\mu(t)} = r(t) - \rho$$

which says the household consumes so the rate of change in consumption is equal to the forgone income she could earn if she invested it in the risk free asset earning return $r(t) = r^k(t) - \delta$.

Production Represent the manufacturing technology by the Cobb-Douglas production function

$$Y_m = \Psi_m K_m^{\alpha_{m1}} L_m^{\alpha_{m2}} \left(Y_{a1}^d\right)^{\alpha_{m3}} \tag{1}$$

The variables in this function are the input levels $K_m(t)$, $L_m(t)$ and $Y_{a1}^d(t)$, while the parameters are Ψ_j, α_{m1}, α_{m2} and α_{m3}, where $\alpha_{m1} + \alpha_{m2} + \alpha_{m3} = 1$. Here, Ψ_j is a "technology" parameter, while α_{m1}, α_{m2} and α_{m3} are the respective factor cost-shares (elasticities) of capital, labor and cotton employed in manufacturing. The cost function corresponding production technology (1) is defined as

$$C^m\left(r^k, w, p_c\right)Y_m \equiv \min_{K_m, L_m, Y_{a1}^d} \left\{r^k K_m + wL_m + p_c Y_{a1}^d : Y_m \le \Psi_m K_m^{\alpha_{m1}} L_m^{\alpha_{m2}} \left(Y_{a1}^d\right)^{\alpha_{m3}}\right\}$$

The cost function is the minimum cost of producing Y_m units of output given factor prices r^k, w and p_c. Given the properties of the Cobb-Douglas function in equation (1), the cost function is twice (continuously) differentiable, non-decreasing and strictly concave in factor prices and increasing in output.

The service sector technology is represented by

$$Y_s = \Psi_s K_s^{\alpha_{s1}} L_s^{1-\alpha_{s1}}$$

where the variable and parameter definitions are analogous to those in equation (1). The cost function associated with the service sector technology is defined as

$$C^s\left(r^k, w\right)Y_s \equiv \min_{K_s, L_s}\left\{r^k K_s + wL_s : Y_s \le \Psi_s K_s^{\alpha_{s1}} L_s^{1-\alpha_{s1}}\right\}$$

As with the manufacturing sector's cost function, $C^s(\cdot)$ is also twice (continuously) differentiable, non-decreasing and strictly concave in factor prices and increasing in output.

Represent the production technology for other agriculture as a function of capital, labor and land, i.e., as

$$Y_{a3} = f^{a3}\left(K_{a3}, L_{a3}, Z_{a3}\right) = \Psi_{a3} K_{a3}^{\alpha_{a31}} L_{a3}^{\alpha_{a32}} Z_{a3}^{\alpha_{a33}} \tag{2}$$

where α_{a31} is the factor share coefficient for capital, α_{a32} is the factor share coefficient for capital labor, and α_{a33} is the factor share coefficient for land, and

$\alpha_{a31} + \alpha_{a32} + \alpha_{a33} = 1$. The value added function corresponding to equation (2) is defined as

$$\Pi^{3a}\left(p_{a3}, r^k, w, Z_{a3}\right) \equiv \max_{K_{a3}, L_{a3}} \left\{ p_{a3} \Psi_{a3} K_{a3}^{\alpha_{a31}} L_{a3}^{\alpha_{a32}} Z_{a3}^{\alpha_{a33}} - r^k K_{a3} - w L_{a3} \right\}$$

For a given technology, the value added function is the maximum land rent that can be generated for a given land endowment and prices p_{a3}, r^k and w. Given the technology (2), the value added function $\Pi^{3a}(.)$ is twice continuously differentiable in prices, increasing in p_{a3}, decreasing in r^k and w, and satisfies Hotelling's lemma.

Cotton and rice production, and other agriculture in South Kazakhstan and Kyzylorda explicitly depend on capital, labor, land and water, with corresponding production technologies:

$$Y_j = \Psi_j K_j^{\alpha_{j1}} L_j^{\alpha_{j2}} Z_j^{\alpha_{j3}} H_j^{\alpha_{j4}}, \quad j = a1, a2, a31, a32 \tag{3}$$

Here, the alpha parameters have the same interpretations as those in equation (2): for example, α_{a11} is the factor share coefficient for capital in cotton production, α_{a12} is the factor share coefficient for capital labor in cotton production, α_{a13} is the factor share coefficient for land in cotton production, and α_{a14} is the factor share coefficient for water in cotton production, with $\alpha_{a11} + \alpha_{a12} + \alpha_{a13} + \alpha_{a14} = 1$. Analogous definitions hold for the rice technology coefficients.

The value-added functions corresponding to (3) are

$$\Pi^j\left(p_j, r^k, w, Z_j, H_j\right) \equiv \max_{K_j, L_j} \left\{ p_j \Psi_j K_j^{\alpha_{j1}} L_j^{\alpha_{j2}} Z_j^{\alpha_{j3}} Z_j^{\alpha_{j4}} - r^k K_j - w L_j \right\},$$

$$j = a1, a2 \tag{4}$$

Equation (4) is the maximum rent farmers can earn on the natural assets land and water: i.e., it is land and water's contribution to GDP.

Equilibrium: The discussion here follows closely that of Roe, Smith and Saraçoğlu (2010).

Given an initial endowment of resources and non-traded good price, $\{K(0), L(0), H(0), Z, p_s(0)\}$, exogenous labor force sequence $\{L(t)\}_{t \in [0,\infty)}$ and exogenous prices (p_{a2}, p_{a3}, p_m), a competitive equilibrium is a sequence of time-dependent prices, capital stock, manufactured good and non-traded good levels, and utility indices, $\{w(t), r(t), p_s(t), K(t), Y_m(t), Y_s(t), \mu(t)\}_{t \in [0,\infty)}$, such that: (i) households intertemporally maximize utility, and at each point in time (ii) firms maximize profit, (iii) capital and labor markets clear and (iv) the non-traded good market clears. The appendix discusses equilibrium in more detail.

Characterization of equilibrium: Given initial conditions on endowments and prices and the labor force sequence, equilibrium is satisfied if (at each t):

Zero profit conditions hold

$$C^m\left(r^k, w, p_{a1}\right) = p_m$$
$$C^s\left(r^k, w\right) = p_s$$

Factor markets clear

$$\frac{\partial}{\partial r^k}C^m\left(r^k,w,p_c\right)Y_m+\frac{\partial}{\partial r^k}C^s\left(r^k,w\right)+\sum_{j=a1,a2,a31,a32,a33}\frac{\partial}{\partial r^k}\Pi^j\left(p_j,r^k,w,\cdot\right)=K(t)$$

$$\frac{\partial}{\partial w}C^m\left(r^k,w,p_c\right)Y_m+\frac{\partial}{\partial w}C^s\left(r^k,w\right)+\sum_{j=a1,a2,a31,a32,a33}\frac{\partial}{\partial w}\Pi^j\left(p_j,r^k,w,\cdot\right)=L(t)$$

The service sector market clears

$$\frac{\partial}{\partial p_s}E\left(p_{a2},p_{a3},p_m,p_s,\mu\right)=Y_s$$

and the following two differential equations are jointly satisfied:

$$\dot{K}=wL+rK+\Pi^{a3}\left(p_{a3},r,w,Z_{a3}\right)+\sum_{j=a1,a2,a31,a33}\Pi^{ja}\left(p_a,r^k,w,Z_j,H_j\right)$$

$$-E\left(p_{a2},p_{a3},p_w,p_s,u\right)$$

$$\dot{p}_s=G\left(K,p_s\right)$$

where $G\left(K,p_s\right)$ is derived as discussed in Roe, Smith and Saraçoğlu (2010), Chapter 4.

The stock and flow values of water: Although not stressed in the earlier discussion, the variable p_h is the price farmers pay for a unit of water. Given there is not market for water, in equilibrium, p_h is the shadow value of an additional unit of water: the amount a farmer would be willing to pay for an additional unit of water. Also, given that p_h embeds the cost of capital, electricity and aquifer rent, it represents the unit gross value of water in agricultural production.

Let $P^h\left(t\right)$ represent the shadow "stock price" of water – the amount a farmer would pay to own the water, and let $P^z\left(t\right)$ represent the purchase price (not rental rate) of land. Given the natural asset stocks Z and $\bar{H}$, the total value of physical and natural asset holdings, denoted $A(t)$ is expressed as

$$A\left(t\right)=K\left(t\right)+P^z\left(t\right)Z+P^h\left(t\right)\bar{H}\left(t\right)$$

Earlier, we noted $\bar{H}\left(t\right)$ represents the period t stock of water. In the empirical application, $\bar{H}$ is the "economically accessible" stock of water, which is defined as $S\left(t\right)=H_0-\int_0^t Y_h\left(t\right)dt$, where $H_0=\int_0^T Y_h\left(t\right)dt$, with T being some period sufficiently in the future: in the empirical model used here, $T=300$.

Assume the natural and physical asset markets are not segmented, and that arbitraging occurs for both types of assets. In such a case, Roe, Smith and

Saraçoğlu (2010) derive the following no-arbitrage condition between r^k and land rents:

$$r^k = \frac{\Pi^a}{P^z} + \frac{\dot{P}^z}{P^z}$$

where $\Pi^a(\cdot, t)$ is time-t agricultural land rent. Smith (2013) derives the following no-arbitrage condition between r^k and the water rent, here interpreted as the gross value of water in agricultural production:

$$r^k = \frac{p_h}{P^h} + \frac{\dot{P}^h}{P^h} + \frac{\dot{\bar{H}}}{\bar{H}}$$

In this case, if arbitrage conditions hold across natural and physical assets, the time t unit stock price of land is given by

$$P^z(t) = \int_t^\infty e^{-\int_t^\vartheta [r^k(v) - \delta] dv} \Pi^a(\cdot, t) dt$$

Here $\Pi^a(\cdot, t) = \sum_{j=1i=1}^{2}\sum^{3}\Pi^{aij}(\cdot)$ is the total land rent for India (Punjab and ROI, rice, wheat and other agricultural land rent). The time-t unit stock price of water is given by

$$P^h(t) = \int_t^\infty e^{-\int_t^\vartheta \left[r^k(v) - \delta - \frac{\dot{\bar{H}}}{\bar{H}}\right] dv} p_h(t) dt \tag{5}$$

Here, the expression $\dfrac{\dot{\bar{H}}}{\bar{H}}$, captures the impact of a declining aquifer on its stock price. If negative, then the effective discount rate

$$e^{-\int_t^\vartheta \left[r^k(v) - \delta - \frac{\dot{\bar{H}}}{\bar{H}}\right] dv} \tag{6}$$

increases, reflecting the loss in value associated with aquifer depreciation. This effect, of course, places a downward pressure on the value of the aquifer.

Appendix 4.2
Other tables

Base scenario

Table 4A.1 Sector share in GDP

Year	Agriculture	Industry	Services
2007	0.0718	0.4932	0.4350
2012	0.0575	0.5035	0.4390
2017	0.0493	0.5087	0.4420
2022	0.0442	0.5116	0.4442
2027	0.0409	0.5132	0.4459
2032	0.0387	0.5143	0.4470
2037	0.0372	0.5149	0.4479
2042	0.0362	0.5153	0.4485
2047	0.0355	0.5156	0.4489
2052	0.0350	0.5158	0.4492

Table 4A.2 Value-added from land and water, and its share in GDP (in US$1,000)

Year	Value added from land and water				GDP	Land and water's share in GDP
	South Kazakhstan	Kyzylorda	ROK	Total		
2007	165.8	37.6	1,664.9	1,868.3	93,435.5	0.0200
2012	166.1	37.9	1,668.8	1,872.8	116,889.0	0.0160
2017	173.7	39.7	1,745.0	1,958.5	142,683.4	0.0137
2022	187.1	42.8	1,879.1	2,109.0	171,315.8	0.0123
2027	205.6	47.2	2,065.4	2,318.2	203,375.5	0.0114
2032	229.3	52.6	2,303.2	2,585.1	239,539.6	0.0108
2037	258.3	59.3	2,594.6	2,912.2	280,578.4	0.0104
2042	293.1	67.4	2,944.0	3,304.5	327,366.5	0.0101
2047	334.3	76.9	3,357.9	3,769.1	380,898.4	0.0099
2052	382.7	88.0	3,844.0	4,314.7	442,309.0	0.0098

Table 4A.3 Ratio of land/water to physical capital stock values

Year	Land and Water (1000 US $)	Capital Stock (1000 US $)	Ratio
2007	43,017,302	252,503,931	0.1704
2012	51,035,116	334,656,067	0.1525
2017	60,217,976	424,544,361	0.1418
2022	70,695,820	523,348,040	0.1351
2027	82,656,046	632,770,755	0.1306
2032	96,332,676	754,938,553	0.1276
2037	112,003,641	892,356,804	0.1255
2042	129,992,505	1,047,911,576	0.1240
2047	150,672,939	1,224,896,942	0.1230
2052	174,475,186	1,427,067,873	0.1223
2057	201,894,188	1,658,710,739	0.1217

Market scenario

Table 4A.4 The value of agricultural production for each sector (in US$1,000)

Year	Other-Ag SK	Other-Ag Kyzylorda	Cotton	Rice
2007	690,902	98,165	58,395	31
2012	693,073	98,474	58,592	35
2017	725,113	103,026	61,310	39
2022	781,100	110,981	66,051	44
2027	858,740	122,012	72,622	50
2032	957,733	136,078	80,998	57
2037	1,079,024	153,311	91,259	65
2042	1,224,460	173,975	103,562	75
2047	1,396,648	198,440	118,128	86
2052	1,598,897	227,176	135,236	99
2057	1,835,237	260,756	155,227	114

Table 4A.5 The stock value of water and land (in US$1,000)

Year	Stock value of water					Stock value of land				
	SKOA	KOA	ROKA	Cotton	Rice	SKOA	KOA	ROKA	Cotton	Rice
2007	348,421	155,371	9,111,272	183,645	672,548	3,386,048	481,100	29,234,660	222,411	45
2012	413,318	184,311	10,808,236	217,851	797,818	4,016,683	570,702	34,679,582	263,857	56
2017	487,638	217,453	12,751,609	257,024	941,276	4,738,891	673,315	40,915,135	311,316	67
2022	572,439	255,268	14,969,041	301,720	1,104,964	5,562,948	790,400	48,030,045	365,464	80
2027	669,239	298,434	17,500,273	352,742	1,291,816	6,503,624	924,054	56,151,819	427,272	94
2032	779,935	347,797	20,394,874	411,087	1,505,490	7,579,338	1,076,895	65,439,511	497,952	111
2037	906,777	404,359	23,711,691	477,943	1,750,331	8,811,961	1,252,029	76,081,933	578,940	129
2042	1,052,385	469,290	27,519,220	554,690	2,031,394	10,226,946	1,453,075	88,298,865	671,909	150
2047	1,219,784	543,938	31,896,556	642,922	2,354,519	11,853,689	1,684,207	102,344,098	778,790	175
2052	1,412,455	629,857	36,934,772	744,475	2,726,429	13,726,034	1,950,235	118,509,844	901,807	203
2057	1,634,407	728,831	42,738,631	861,461	3,154,856	15,882,915	2,256,691	137,132,253	1,043,518	235

Table 44.6 Water and land wealth in the water market and base scenarios (in US$1,000)

Year	Market Scenario			Base Scenario			Difference		
	Water	Land	Total	Water	Land	Total	Water	Land	Total
2007	10,436,037	32,581,271	43,017,308	10,279,294	32,691,971	42,971,265	-156,743	110,700	-46,043
2012	12,382,151	38,652,974	51,035,124	12,216,335	38,806,993	51,023,328	-165,815	154,019	-11,796
2017	14,610,835	45,607,150	60,217,985	14,446,075	45,816,223	60,262,297	-164,761	209,073	44,312
2022	17,153,669	53,542,162	70,695,832	17,003,280	53,821,401	70,824,681	-150,389	279,239	128,850
2027	20,056,152	62,599,908	82,656,060	19,937,100	62,967,051	82,904,151	-119,052	367,143	248,091
2032	23,375,087	72,957,604	96,332,692	23,308,956	73,432,802	96,741,758	-66,131	475,198	409,066
2037	27,177,929	84,825,729	112,003,659	27,192,303	85,431,449	112,623,752	14,374	605,720	620,093
2042	31,543,201	98,449,324	129,992,526	31,673,472	99,210,083	130,883,555	130,270	760,759	891,029
2047	36,561,583	114,111,380	150,672,963	36,853,207	115,052,992	151,906,199	291,623	941,612	1,233,236
2052	42,337,482	132,137,733	174,475,214	42,848,716	133,285,584	176,134,299	511,234	1,147,851	1,659,085
2057	48,991,002	152,903,219	201,894,220	49,796,098	154,278,629	204,074,727	805,097	1,375,410	2,180,507

Table 4A.7 Ratio of land/water to physical capital stock values

Year	Land and Water (US$1,000)	Capital Stock (US$1,000)	Ratio
2007	43,795,522	252,503,931	0.1734
2012	51,952,415	334,535,498	0.1553
2017	61,293,724	424,277,313	0.1445
2022	71,952,369	522,915,548	0.1376
2027	84,119,367	632,156,506	0.1331
2032	98,032,990	754,126,079	0.1300
2037	113,976,095	891,327,617	0.1279
2042	132,277,925	1,046,643,848	0.1264
2047	153,318,680	1,223,364,500	0.1253
2052	177,536,112	1,425,239,323	0.1246
2057	205,433,798	1,656,548,608	0.1240

Improving irrigation efficiency

Table 4A.8 Sector value-added with improved irrigation efficiency (in US$1,000)

Year	SKOA	KOA	ROKA	Cotton	Rice	Industry	Services
2007	531,162	99,036	5,967,396	62,718	30,394	46,487,453	40,499,619
2012	539,024	100,502	5,997,270	64,105	32,085	59,142,215	51,101,801
2017	569,851	106,250	6,281,450	68,237	35,098	72,776,356	62,791,404
2022	619,768	115,557	6,770,688	74,703	39,345	87,731,015	75,773,297
2027	687,505	128,187	7,446,062	83,392	44,855	104,356,120	90,302,924
2032	773,265	144,177	8,305,353	94,367	51,732	123,026,901	106,681,047
2037	878,206	163,743	9,356,815	107,804	60,138	144,158,224	125,691,588
2042	1,004,203	187,236	10,616,212	123,971	70,289	168,335,589	146,843,550
2047	1,153,744	215,118	12,105,455	143,216	82,449	195,969,794	171,017,042
2052	1,329,884	247,960	13,851,923	165,959	96,935	227,650,133	198,724,189
2057	1,536,198	286,427	15,887,776	192,696	114,118	264,048,582	230,550,920

Table 4A.9 Sector share in GDP

Year	Agriculture	Industry	Services
2007	0.0711	0.4962	0.4323
2012	0.0573	0.5056	0.4369
2017	0.0493	0.5103	0.4402
2022	0.0443	0.5127	0.4428
2027	0.0411	0.5139	0.4447
2032	0.0390	0.5146	0.4462
2037	0.0375	0.5149	0.4474
2042	0.0365	0.5150	0.4483
2047	0.0358	0.5150	0.4489
2052	0.0353	0.5151	0.4494
2057	0.0350	0.5152	0.4496

Table 4A.10 Water flow shadow values – per unit and total

Year	Unit shadow rental rate per sector ($/km³)					Shadow value-added per sector (US$1,000)					Total
	SKOA	KOA	ROKA	Cotton	Rice	SKOA	KOA	ROKA	Cotton	Rice	
2007	48.80	20.41	70.85	18.93	6.08	35,140	6,552	394,787	7,183	8,454	452,117
2012	48.92	20.45	71.01	18.97	6.19	35,660	6,649	396,763	7,342	8,924	455,339
2017	51.15	21.39	74.26	19.84	6.54	37,700	7,029	415,564	7,815	9,762	477,871
2022	55.08	23.03	79.96	21.37	7.09	41,002	7,645	447,931	8,556	10,944	516,078
2027	60.54	25.31	87.89	23.49	7.83	45,483	8,480	492,612	9,551	12,476	568,603
2032	67.51	28.23	98.01	26.19	8.77	51,157	9,538	549,460	10,808	14,389	635,353
2037	76.05	31.80	110.41	29.51	9.90	58,100	10,833	619,022	12,347	16,727	717,029
2042	86.30	36.08	125.28	33.48	11.26	66,435	12,387	702,340	14,199	19,551	814,912
2047	98.43	41.16	142.89	38.19	12.86	76,329	14,232	800,864	16,403	22,933	930,760
2052	112.68	47.11	163.57	43.72	14.74	87,982	16,404	916,406	19,008	26,962	1,066,762
2057	129.33	54.08	187.75	50.18	16.93	101,631	18,949	1,051,093	22,070	31,741	1,225,484

Table 4A.11 Unit stock shadow price of water

Year	"Correct" calculations					Price/earnings ratio				
	SKOA	KOA	ROKA	Cotton	Rice	SKOA	KOA	ROKA	Cotton	Rice
2007	982.7	410.9	1,628.5	392.5	140.6	624.4	261.1	906.5	242.2	77.8
2012	1,133.3	473.9	1,931.2	455.6	166.8	674.3	281.9	1,009.7	263.4	87.4
2017	1,306.3	546.2	2,278.1	528.0	196.6	739.8	309.3	1,141.5	291.0	99.2
2022	1,503.7	628.7	2,674.3	610.6	230.5	819.6	342.7	1,301.5	324.5	113.3
2027	1,729.3	723.1	3,126.9	705.0	269.1	913.5	382.0	1,491.6	364.0	129.8
2032	1,988.0	831.2	3,644.8	813.0	313.2	1,022.4	427.5	1,714.5	409.8	149.0
2037	2,285.8	955.7	4,238.6	937.3	363.7	1,147.3	479.7	1,974.3	462.6	171.1
2042	2,630.0	1,099.7	4,920.7	1,080.7	421.8	1,290.0	539.4	2,276.2	523.1	196.7
2047	3,029.6	1,266.7	5,705.1	1,246.7	488.6	1,452.7	607.4	2,626.0	592.3	226.1
2052	3,495.2	1,461.4	6,608.2	1,439.6	565.6	1,637.9	684.8	3,031.0	671.3	260.0
2057	4,040.1	1,689.2	7,648.6	1,664.6	654.7	1,848.5	772.9	3,499.5	761.6	299.1

Table 4A.12 The stock value of water and land

Year	Stock value of water (US$/km³)					Stock value of land				
	SKOA	KOA	ROKA	Cotton	Rice	SKOA	KOA	ROKA	Cotton	Rice
2007	709,045	132,203	9,092,912	149,268	195,866	2,711,397	505,546	29,175,746	256,809	42,474
2012	851,785	158,817	10,783,202	180,492	242,038	3,242,209	604,517	34,599,256	308,914	52,097
2017	1,021,009	190,369	12,720,470	217,530	296,697	3,851,089	718,044	40,815,221	368,733	63,136
2022	1,220,541	227,573	14,932,733	261,236	361,197	4,547,094	847,815	47,913,545	437,174	75,773
2027	1,455,681	271,415	17,459,952	312,770	437,283	5,342,739	996,165	56,022,442	515,465	90,239
2032	1,733,201	323,159	20,351,897	373,604	527,095	6,253,346	1,165,950	65,301,612	605,092	106,802
2037	2,061,525	384,376	23,667,639	445,552	633,211	7,296,821	1,360,508	75,940,585	707,772	125,764
2042	2,451,028	456,999	27,475,904	530,831	758,709	8,493,653	1,583,660	88,159,881	825,443	147,447
2047	2,914,458	543,407	31,855,948	632,143	907,250	9,867,006	1,839,724	102,213,801	960,262	172,199
2052	3,467,494	646,522	36,898,738	752,782	1,083,180	11,442,845	2,133,543	118,394,223	1,114,601	200,371
2057	4,129,468	769,948	42,708,264	896,772	1,291,647	13,249,993	2,470,490	137,034,815	1,291,019	232,312

Table 4A.13 Ratio of land/water to physical capital stock values

Year	Land and Water (US$1,000)	Capital Stock (US$1,000)	Ratio
2007	42,971,265	253,771,696	0.1693
2012	51,023,328	335,202,827	0.1522
2017	60,262,297	424,366,435	0.1420
2022	70,824,681	522,434,534	0.1356
2027	82,904,151	631,097,549	0.1314
2032	96,741,758	752,470,586	0.1286
2037	112,623,752	889,059,745	0.1267
2042	130,883,555	1,043,777,441	0.1254
2047	151,906,199	1,219,994,169	0.1245
2052	176,134,299	1,421,635,639	0.1239
2057	204,074,727	1,653,373,812	0.1234

Notes

1 Using four different model specifications, Gemma and Smith (2019) estimate the equilibrium levels of five variables: the price of two types of rice, the land rental rates associated with each type of rice produced (identical in equilibrium) and the number of production quotas traded. The model specifications involve: Cobb-Douglas production and utility, Cobb-Douglas production and Almost-Ideal-Demand-System (AIDS) utility, "quasi-Leontief" production and Cobb-Douglas utility functions, and "quasi-Leontief" production and AIDS utility. They find each model yields close to identical rice prices and land rent, but quite different quota trading levels.
2 Another policy experiment not implemented here would be to introduce a water market, but decrease the amount of water to trade enough to keep total asset wealth at least as large as the status quo outcome.
3 Asset wealth is non-declining over time.
4 The model for the third scenario (improved irrigation efficiency) is almost identical to the model presented in this section. Modelling the second scenario (water trading) requires a few changes in the modelling setup: water becomes a choice variable in Syr Darya agriculture's production functions, and one needs to introduce a market clearing condition for water.
5 Land is fixed at each point in time.

References

Bekchanov, M., A. Bhaduri, and C. Ringler. "Potential Gains from Water Rights Trading in the Aral Sea Basin." *Agricultural Water Management*, Vol. 152 (2015), 41–56.

Gemma, M., and R.B.W. Smith. "The Economics of Rice Production Quota Trading." Unpublished working paper, available upon request, 2019.

Ghany, A., and A. Shawky. *Project Information Document (Concept Stage): Syr Darya Control and Northern Aral Sea Project, Phase 2: P152001*. World Bank, Washington, DC, 2014. http://documents.worldbank.org/curated/en/2014/09/20234171/project-information-document-concept-stage-syr-darya-control-northern-aral-sea-project-phase-2-p152001

Japan Aerospace Exploration Agency, www.eorc.jaxa.jp/en/imgdata/topics/2007/tp071226. html

Micklin, P. "Introduction to the Aral Sea and Its Region." In *The Aral Sea: The Devastation and Partial Rehabilitation of a Great Lake*, edited by P. Micklin, N.V. Aladin and I. Plotnikov. Springer, New York, 2014.

Micklin, P., N.V. Aladin, and I. Plotnikov. *The Aral Sea: The Devastation and Partial Rehabilitation of a Great Lake*. Springer, New York, 2014.

Murray-Rust, H., I. Abdullaev, ul M. Hassan, and V. Horinkova. *Water Productivity in the Syr-Darya River Basin*. Research Report 67. International Water Management Institute, Colombo, Sri Lanka, 2003.

Regional Environmental Center for Central Asia (CAREC). From tables provided by CAREC staff, 2015. English interpretations of data from Kazakhstan Ministry of Agriculture.

Roe, T.L., R.B.W. Smith, and S. Saraçoğlu. *Multisector Economic Growth Models: Theory and Applications*. Springer, New York, 2010.

Roland-Holst, D. and Kazybayeva, S. Social Accounting Matrix for Kazakhstan. GTAP 7, Data Base Documentation, www.gtap.org, 2009.

Smith, R.B.W., H. Nelson, and T.L. Roe. "Groundwater and Economic Dynamics, Shadow Rents and Shadow Prices in General Equilibrium Settings." *Water Economics and Policy*, Vol. 1, Issue 03, September 2015.

Smith, R.B.W. "Ecosystem Services and the Macroeconomy: A Review of Linkages and Evaluation of Analytical Tools." United Nations Environment Program, Ecosystem Services Economics, Working Paper Series, Division of Environmental Policy Implementation, Paper No 20, September 2013.

Smith, R.B.W., and M. Gemma. "Valuing Ecosystem Services in Macroeconomic Settings." In *Handbook on the Economics of Ecosystem Services and Biodiversity*, edited by P. Nunes, P. Kumar and T. Dedeurwaerdere. Edwin Elgar, Cheltenham, UK, June 2014.

Smith, R.B.W., H. Nelson, and T.L. Roe. "Groundwater and Economic Dynamics, Shadow Rents and Shadow Prices in General Equilibrium Settings." *Water Economics and Policy*, Vol. 1, Issue 03, September 2015.

World Bank. "World Development Indicators." https://databank.worldbank.org/reports. aspx?source=world-development-indicators, 2015.

5 Mapping the cultural services of ecosystems and heritage sites in the Usumacinta floodplain in Mexico

Andrea Ghermandi and
Vera Camacho-Valdez

Introduction

Tourism, recreation and other direct interactions with the natural and cultural environment may play an important role in the conservation and sustainable management of natural ecosystems and cultural heritage sites. The main argument in support of ecotourism (or nature-based tourism) and cultural heritage tourism – which is understood here as tourism encompassing activities that revolve around built patrimony, living lifestyles, ancient artifacts, and modern art and culture (Dallen 2011) – is that it promotes win-win scenarios, in which tourists benefit from an enjoyable experience, there are economic benefits for tour operators, and parts of the funds raised are reinvested in environmental conservation and improving livelihoods within local communities (Ardoin et al. 2015; Stronza and Durham 2008). While the debate on the extent to which such benefits are actually realized (Higham 2007; Torre and Scarborough 2017) and what constitutes sustainable tourism (Asmelash and Kumar 2019) are still ongoing, the growth of the (eco)tourism industry in many parts of the world makes this an important area of research, especially where destination sites are ecologically and/or culturally fragile (Balmford et al. 2015).

In this context, previous research has shown that local, domestic and international visitors often differ in their cultural preferences as well as the spatial distribution of their visitation patterns. Visits by tourists tend to be more spatially concentrated than those by residents (Garcia-Palomares et al. 2015; Munoz et al. 2019), focusing on hotspots that are more easily accessible and better equipped with infrastructure (Heagney et al. 2017; Su et al. 2016). Domestic visitors may hold different attitudes toward wilderness areas compared to international visitors, whereby the expectations of the latter are often informed by the marketing strategies of tour operators (Higham et al. 2001). Place-based values within natural areas that are mapped by local residents in the context of participatory GIS studies tend to differ in their spatial distribution from those produced by domestic and international visitors and, even where they overlap, locals and tourists may associate different benefits for the same areas (Munoz et al. 2019; Munro et al. 2017).

Understanding such differences may have important management implications in particular for areas that include multiple attractions, such as sites of natural and cultural interest. Characterizing the similarities and dissimilarities in the way different types of visitors interact with them may, for instance, help avoid conflicts in areas where increased international tourism threatens the enjoyment of cultural benefits held by local residents (Wray et al. 2010). This is especially important in the context of the rising anti-tourism sentiment that has recently been experienced in various popular tourist destinations (Coldwell 2017; Clancy 2017; Mihalic 2018). Moreover, such information can be used to optimize the location of infrastructure and services provided to the visitors, as well as provide business opportunities for meeting their specific demands and developing currently touristically unexploited areas (Garcia-Palomares et al. 2015).

The digital revolution offers opportunities to improve our understanding of how ecosystems and cultural heritage sites are engaged with or enjoyed by visitors. Traditionally, such information is collected through surveys, interviews (whether on-site, phone- or Internet-based) or focus groups with destination tourism stakeholders, which are however time- and resource-intensive and often impractical at large scales. Recently, however, a new generation of geolocated information sources has emerged, which revolves around various forms of crowdsourcing, i.e., the collection and sharing of data by non-professionals and citizen organizations (Conrad and Hilchey 2011). Such techniques often require an active engagement on the part of the "neogeographers" (Goodchild 2009) who are involved in providing Volunteered Geographic Information (Elwood et al. 2012), for instance through Public Participatory Geographic Information Systems (Munoz et al. 2019) or in citizen science projects (Tipaldo and Allamano 2017; Velwaert and Caley 2016). Particularly promising in terms of volume of data, large-scale coverage and near real-time data collection are, however, passive crowdsourcing techniques (Ghermandi and Sinclair 2019). These rely on information that is produced either involuntarily (e.g., through interaction with outdoor surveillance cameras or devices that are capable of location tracking such as smartphones) or voluntarily but for purposes other than those which the users originally intended (e.g., analysis of information publicly shared by the users of social networking services) (Connors et al. 2012). Among the latter, the analysis of geotagged photographs from social media sites has gathered particular attention among environmental researchers (Ghermandi and Sinclair 2019).

In this study we develop and apply techniques to investigate the spatial distribution of cultural ecosystem services and historical heritage tourism and recreation activities, as enjoyed by either local, domestic or international visitors to the region of the Usumacinta floodplain, a 25,000 km^2 coastal region with one of the highest biological and cultural diversities in Mexico. For this purpose, we analyze geotagged photographs uploaded by visitors to the popular photo-sharing website Flickr (www.flickr.com).

Such analysis is motivated by the fact that while several previous studies have investigated how online photographs can be used to identify specific cultural services (e.g., Donaire et al. 2014) or the spatial distribution of visitors based on their origin (e.g., Garcia-Palomares et al. 2015), research on how specific cultural services are accrued to different users, including their spatial dimension, is largely missing. One exception is a recent investigation of the preferences held by local, regional and international visitors for ecosystem benefits in Great Lakes Areas of Concern (Angradi et al. 2018), which will be further discussed in the following sections.

The remainder of this chapter is organized as follows. The section "Social media and cultural ecosystem services" summarizes the background information regarding the use of social media data and, in particular, geotagged photographs in the analysis of cultural ecosystem services and cultural heritage tourism. The section titled "Case-study area: the Usumacinta floodplain" presents the case-study area. The section "Materials and methods" discusses the methodologies implemented in the present study for the retrieval of geotagged photographs' data, spatial analysis of the distribution of photos, analysis of the origin of the users who uploaded the photographs, classification of the photographs based on the type of cultural service they reflect, and statistical analyses. The sections "Results" and "Discussion and conclusions" respectively summarize the main results of the study and discuss them in the context of the literature and for their implication for the management of the natural capital and cultural heritage sites in the case-study region.

Social media and cultural ecosystem services

According to a systematic review of the applications of passively crowdsourced data from social media in environmental research (Ghermandi and Sinclair 2019), the most frequent application for this data is the analysis of cultural ecosystem services, i.e., the physical/experiential, intellectual/representative or spiritual/ symbolic interactions people have with the natural environment (Haines-Young and Potschin 2018). These studies primarily focus on the characterization of non-extractive recreational activities (e.g., hiking, walking, birdwatching, boating), including temporal and spatial patterns, and the aesthetic value of landscapes. Among the applications that are of greatest relevance for the present study, one may count the evaluation of factors contributing to eco-tourist satisfaction (Tenkanen et al. 2017; Hausmann et al. 2017), the extraction of points of interest or hot spots of cultural value (Figueroa-Alfaro and Tang 2017; Ghermandi 2016; Lee et al. 2014; Levin et al. 2015), and the mapping of aesthetic appreciation of landscapes and scenic areas (van Zanten et al. 2016; Langemeyer et al. 2018; Lee et al. 2019; Van Berkel et al. 2018). The investigation of online photographs from photo-sharing websites (such as Flickr, Instagram and Panoramio) and text from microblogging platforms (such as Twitter and Weibo) are the two most common applications of social media in environmental research, with Flickr being second only to Twitter in the number of published studies (Ghermandi and Sinclair 2019). Geotagged photo counts generally show a good correlation with observed spatial and temporal patterns of visitation (Wood et al. 2013; Sinclair et al. 2018; Tenkanen et al. 2017).

Previous studies have used a range of techniques for distinguishing between photographs uploaded by local residents, domestic tourists and international tourists. The analysis of the information provided by Flickr users in their profiles, may be used to determine their hometown or current location, but only about 40–48% of the users provide this information (Tenerelli et al. 2017; Da Rugna et al. 2012; Vu et al. 2015).

The username and/or the photographer's attitude towards the camera (e.g., whether he/she conforms to the convention of standing and smiling in front of the camera) may also provide useful indications concerning the user's provenance or whether he/she is a tourist (Angradi et al. 2018; Donaire et al. 2014). Some studies propose to use time-based approaches such as a minimum time interval between photographs (Li et al. 2013) or user activity restricted to a narrow timeframe over prolonged periods of time (Straumann et al. 2014) to distinguish between local residents and tourists. Rules based on activity time span were used for instance by Koerbitz et al. (2013) and Garcia-Palomares et al. (2015). Da Rugna et al. (2012) propose a combination of learning algorithms and expert-defined rules relying on time-based criteria to identify the country of origin of Flickr users. Bojic et al. (2015) propose five methods to infer the home location of Flickr and Twitter users, which rely on determining the place where the user took the maximal number of photos, spent the maximal number of user days (i.e., days in which at least one photo was taken), where the time span between the first and last photograph is maximal, where the user took the maximal number of photos or was most active during night hours. Ghermandi (2018) and Sinclair et al. (2018) explored several of the techniques proposed by Bojic et al. (2015) and Li et al. (2013), concluding that the most accurate results are provided by the rule that infers home location from the place with most active user days.

The three main approaches that have been proposed in the literature to associate online photographs with the specific type(s) of cultural services they reflect rely on the analysis of: (1) the content of the photographs; (2) the text associated with photographs' titles and tags and (3) a combination of the two. The most common approach consists in manually analyzing the actual content of individual photos in order to classify them into categories (e.g., "Nature", "Heritage", "Culture" and "Tourist services") based on the presence or absence of specific elements in the photos, such as views of flora and fauna, historical buildings, or tourist infrastructure and facilities (Donaire et al. 2014; Martinez-Pastur et al. 2016; Tieskens et al. 2018; Heikinheimo et al. 2017). Some authors rely on categories that do not reflect common classifications of cultural ecosystem services to avoid investigator biases related to the perceived subjective nature of such classifications (Oteros-Rozas et al. 2018; van Berkel et al. 2018). More nuanced, but potentially more open to subjective interpretations, versions of this approach involve trying to account for the intent of the photographer when taking the photograph (Angradi et al. 2018) or focusing on the main subject of the photograph only (Casalegno et al. 2013; Hausmann et al. 2017; Richards and Friess 2015). A different approach relies on the investigation of the text associated with the titles and tags of the photographs. The presence of keywords may be used to identify specific types of cultural ecosystem services (Spalding et al. 2017; Mancini et al. 2018; van Zanten et al. 2016) or identify and eliminate irrelevant photographs (Mancini et al. 2018). Lists of keywords may be defined a priori by the investigators (van Zanten et al. 2016) or built bottom-up from the analysis of (a sample of) the entire corpus of keywords in the set of photographs under investigation (Mancini et al. 2018; Dunkel 2015). A third approach consists in relying on machine learning algorithms to automatically tag photos based on the content of the image and subsequently use such tags to classify the photographs into categories of cultural ecosystem services (Richards and Tuncer 2018).

Case-study area: the Usumacinta floodplain

The Usumacinta floodplain (Figure 5.1) is located in the southern Gulf of Mexico and is considered to be among the richest in Mexico for natural capital and cultural heritage (Hudson et al. 2005; Carabias et al. 2010). Freshwater pulses with high suspended sediments and inorganic nutrients and organic materials generate

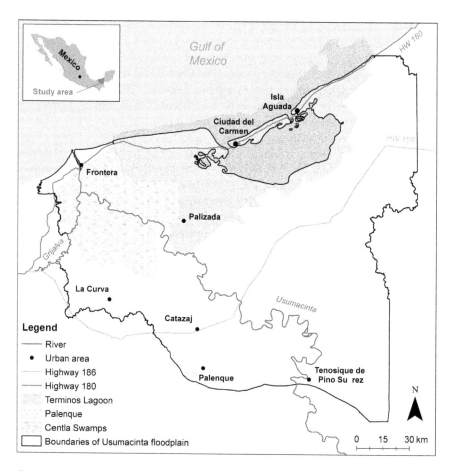

Figure 5.1 Study area: the Usumacinta floodplain in Southern Mexico. The river and highway layers are derived from OpenStreetMap data.

extensive wetlands (e.g. mangroves and coastal lagoons), notably including Terminos Lagoon and Centla Swamps. These two natural protected areas constitute major portions of the floodplain system (Yáñez-Arancibia et al. 2009), support a substantial fishing activity for local communities and maintain high diversity of invertebrates and aquatic vertebrates representative of the tropical wetlands of Mesoamerica (Sanchez et al. 2012). Fisheries also include reef fish, coastal migratory pelagic fish and large oceanic pelagics of great importance at an international level (Yañez-Arancibia and Day 2004), which also depend on the ecological integrity of the Usumacinta floodplain system, its waters and the quality of their habitats.

Given the richness of their ecosystems together with its cultural heritage sites, such as the Mayan ruins at Palenque, this region has become a very attractive destination for tourism and recreation activities. The archaeological zone of Palenque alone received around 600,000 visitors during 2016 (www.estadisticas.inah.gob.mx/). Tourism is an important source of foreign currency in the region. Other important economic activities in the region are oil and gas production, and agriculture. These activities often lead to contamination and habitat destruction, inducing uncertainty both in economic development but also leading to conflicts of interest with environmental values (Yáñez-Arancibia 1999).

The importance of nature-based and cultural heritage tourism and recreation for both local residents and tourists, the variety of natural ecosystems that are present (coastal lagoons, mangroves, tropical forest, sandy beaches, riverine ecosystems), as well as the threats from pollution and habitat destruction that it is currently experiencing, make the Usumacinta floodplain an ideal region in which to test the applicability of social media data analysis for the mapping and characterization of cultural services.

Materials and methods

We retrieved and analyzed the metadata of 8,245 geotagged photographs taken within the boundaries of the Usumacinta floodplain using Flickr's Application Programming Interface (API). The photographs were taken between January 1, 2004, and March 16, 2017, and uploaded to Flickr by 499 individual users. The public profile of all 499 users was investigated to determine the current home location if reported or, in the absence of such information, the hometown of the visitors. For users who do not disclose their current location or hometown in their profile, the metadata of all public photos they uploaded to Flickr were downloaded and analyzed to determine the area with the highest number of active user days, according to the procedure described in Ghermandi (2018) and Sinclair et al. (2018). Users residing within the states of Chiapas, Tabasco or Campeche were classified as local residents. A further distinction was established between domestic tourists residing in other parts of Mexico and international visitors.

For the identification of the cultural services associated with the photos, we relied on the CICES 5.1 classification (Haines-Young and Potschin 2018). Since the study builds on photographs that were taken by actual visitors to the area, we only considered the five classes of cultural ecosystem services that reflect direct, in-site and outdoor interactions that depend on actual presence in the environmental setting. Services reflecting "physical and experiential interactions with the natural environment" were further subdivided based on whether the interaction is with plants, birds, other wild animals, or specific elements of the landscape (e.g., waterfalls). Services associated with historical sites were incorporated in the class of cultural heritage services. No distinction was made between services derived from biotic or abiotic components of the ecosystems.

In order to classify the photographs based on the class of cultural service they reflect, if any, we investigated the text associated with the titles and tags of the photos. A total of 24,517 words were extracted from the 6,317 photographs that were associated with a title and/or one or more tags. Most words were in Spanish and English, but German, French and Italian words were also found, in addition to Latin terms identifying the scientific names of various animal and plant species. After removing duplicates, the words were investigated to identify terms associated with specific cultural services. Table 5.1 provides an overview of the cultural services identified as well as examples of keywords for each of the services. Photographs with multiple keywords in title or tags could be classified under multiple cultural services. The authors jointly reviewed the classification of keywords, and divergences were discussed until an agreement was reached. To test the reliability of the keyword classification procedure, the content of 278 of the photos with a meaningful title and/or tags was independently assessed for whether the photograph was associated with a cultural service. The agreement between the keyword and content-based classifications was evaluated by means of Cohen's kappa coefficient (Cohen 1960).

For the analysis of the spatial distribution of photographs, we relied on the Kernel Density and Hot Spot Analysis (Geit-Ord Gi*) tools with a grid cells size

Table 5.1 Cultural services and examples of keywords associated with them

CICES group	Cultural service	Examples of associated keywords
Physical and experiential interactions with natural environment	Observation of birds	Amazon kingfisher, birding, *Botaurus pinnatus*, heron, mycteriaamericana, . . .
	Observation of other animals	Ameisenhuegel, blackiguana, fantasticwildlife, lizards, spidermonkey, . . .
	Observation of plants	Albero, flores, palmeiras, rainforest, . . .
	Observation of element of landscape	Aguaazul, cascadas, fiume, laguna, pantanos, seascape, waves, . . .
Intellectual and representative interactions with natural environment	Aesthetic and mental health	Absolutelystunningscapes, bellezanatural, paradise, salvaje, tranquilidad, . . .
	Cultural heritage	Ancientruins, archeologicalsite, culturamaya, historisch, maya, Palenque, . . .

Source: author's computation

of 300 m and a radius of 5 km, as implemented in ArcGIS 10.6.1. The tools were used to identify clustering of photographs associated to cultural services while distinguishing between photographs taken by locals, other domestic visitors, or international tourists.

The probability that users of different origins were associated with specific cultural services was explored through logistic regression, controlling for the fact that users who took many photographs within the region were more likely to be associated with culturally tagged photographs and relying on the Wald test to evaluate whether the overall effect of the user's place of origin was statistically significant.

Results

In total, 264 of the investigated Flickr users reported their home location in their profile, corresponding to 53% of the total sample of users, a percentage that is slightly higher than what was found by Da Rugna et al. (2012) and Tenerelli et al. (2017). Through the analysis of the entire corpus of public photos uploaded to Flickr, we could infer the home location of an additional 205 users. Local residents, domestic tourists, and international tourists accounted for 19%, 30% and 51% of the sampled users, respectively. Most of the international tourists were from European countries (27%) and other North American countries (14%).

Of the 6,317 photographs with titles and/or tags, 3,476 were found to be associated with one or more cultural services. Figures 5.2 and 5.3 show the results of the spatial analysis of the distribution of culturally tagged photographs, according to the provenance of the user associated with them.

Culturally tagged photographs taken by international visitors are primarily concentrated in correspondence to the archaeological sites at the Mayan ruins of Palenque, reaching in that area a substantially higher density than those of other visitors. Although the Palenque area appears to be a site of high cultural significance also for locals and domestic visitors, the photographs associated with cultural keywords from local and domestic visitors are more widespread within the floodplain. For these visitors, additional areas of high density of culturally

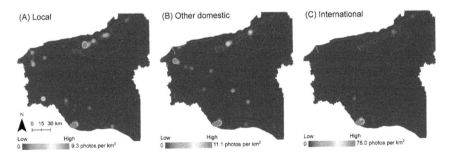

Figure 5.2 Kernel density of photographs associated with at least one cultural service for locals, domestic tourists and international tourists.

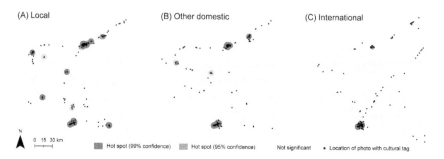

Figure 5.3 Location of photographs associated with at least one cultural service and hot
spots of cultural services for locals, domestic visitors and international tourists.

tagged photographs and hot spots are observed in correspondence of the Terminos Lagoon (e.g., Ciudad del Carmen, Isla Aguada), Centla Swamps (e.g., Usu-
macinta-Grijalva confluence), and other urban areas such as Frontera, Palizada
and, only for local residents, La Curva, Catazajá and Tenosique de Pino Suárez.
For all users there appears to be some correspondence between the location of
photographs and the presence of urban areas and major roads in the region. In
particular, there appears to be an alignment with the Federal Highway 180, which
runs parallel to the coast, and, for international visitors only, with the Federal
Highway 186. The latter observation is consistent with the fact that highway 186
is of high importance for the transit of tourists toward some the most touristic
zones in Mexico located in the Yucatan Peninsula.

The analysis of the photographs' titles and tags revealed 813 individual keywords that could be associated with cultural ecosystem services and cultural
heritage tourism. Comparison of such keyword-based classification with the
classification of photographs based on the image content reveals a fair-to-good
agreement between the methods. For the test subset of 278 photographs with a
meaningful description in their titles or tags, we found an 87.5% overall classification agreement (Cohen's kappa = 0.54). On average, international tourists
uploaded more photographs (15.3 photos per capita) than domestic tourists (7.2
photos per capita) and local residents (12.3 photos per capita). Visitors from the
United States and Canada were particularly active (23.0 photos per capita). The
majority of photographs from international tourists (67%) were associated with
at least one cultural service, compared to 46% and 50% for locals and domestic
tourists, respectively. In particular, 62% of the photos from international tourists
pertained to cultural heritage, with a peak of 72% for European visitors.

Table 5.2 shows the results of the statistical regression analysis. After controlling for the number of culturally tagged photos taken within the study region,
whose sign is as expected positive and statistically significant for all models, the
logistic regression confirms that international tourists are more likely to be associated with cultural photographs among all visitors (N = 468), and in particular with

Table 5.2 Results of logistic regression for visitors' association with cultural services based on origin and number of geotagged photos taken in the Usumacinta Floodplain

	Any cultural service	Physical and experiential: Observation of				Intellectual and representative	
		Birds	Other animals	Plants	Element of landscape	Aesthetic and mental health	Cultural heritage
Intercept: estimate	0.718***	−3.495***	−3.310***	−2.115***	−1.880***	−2.838***	0.467***
[95% confidence interval]	[0.363, 1.084]	[−4.332, −2.768]	[−4.082, −2.633]	[−2.619, −1.650]	[−2.331, −1.458]	[−3.450, −2.285]	[0.126, 0.815]
Photos: estimate	0.566***	0.846***	1.463***	0.919***	0.918***	1.095***	0.431**
[95% confidence interval]	[0.212, 0.934]	[0.291, 1.407]	[0.939, 2.019]	[0.522, 1.325]	[0.562, 1.282]	[0.671, 1.532]	[0.099, 0.769]
Other domestic: estimate	−0.583**	0.287	−1.032**	−0.432	0.324	0.164	−1.022***
[95% confidence interval]	[−1.035, −0.132]	[−0.631, 1.162]	[−2.149, −0.109]	[−1.067, 0.166]	[−0.178, 0.823]	[−0.509, 0.814]	[−1.460, −0.590]
Local: estimate	−0.532**	1.045**	−0.200	−0.206	0.159	1.049***	−1.608***
[95% confidence interval]	[−1.057, −0.001]	[0.201, 1.881]	[−1.123, 0.629]	[−0.905, 0.445]	[−0.440, 0.738]	[0.405, 1.692]	[−2.158, −1.083]
Degrees of freedom	468	468	468	468	468	468	468
Null deviance	587.88	248.99	282.81	425.30	524.71	398.75	648.62
Residual deviance	567.96	234.96	242.43	400.39	498.43	363.31	593.21
Residual deviance test: p-value	<0.001	0.003	<0.001	<0.001	<0.001	<0.001	<0.001
Log-likelihood	−283.98	−117.48	−121.21	−200.20	−249.21	−181.81	−296.61
AIC	575.96	242.96	250.43	408.39	506.43	371.61	601.21
Wald test: p-value	0.021	0.044	0.130	0.370	0.440	0.004	<0.001
Error rate	32.0%	7.5%	8.7%	17.1%	25.6%	14.7%	35.0%

Source: Author's computation

Notes: AIC = Akaike Information Criterion;
*** and ** indicate respectively 1% and 5% statistical significance levels (p-value). "International" is the omitted variable for users' origin.

Table 5.3 Probability of visitors being associated with a culturally tagged photo measured at sample mean number of photos

Cultural service	International	Other domestic	Local
Birds	7.1%	9.2%	17.8%
Other animals	15.3%	6.0%	12.8%
Plants	24.7%	17.6%	21.1%
Element of landscape	29.3%	36.5%	32.7%
Aesthetic and mental health	16.2%	18.5%	35.5%
Cultural heritage	71.8%	47.9%	33.8%

Source: author's computation

cultural heritage. Interestingly, though, local visitors are more likely to be associated with birdwatching and photographs reflecting aesthetic value and mental health than any other visitor type. Domestic visitors are less likely to be associated with photographs of wild animals (other than birds) than international tourists and local residents, although the statistical significance of such finding is not confirmed by the Wald test (p = 0.130).

Table 5.3 builds on the results of the logistic regression to evaluate the probability of visitors to be associated with specific types of cultural services. Probabilities in Table 5.3 are calculated at the sample mean number of photos (12.3 photos per capita), thus controlling for the fact that international visitors take and upload more geotagged photographs than locals and domestic tourists. International visitors are overall more likely to be associated with at least one culturally tagged photograph, but this is largely driven by the fact that they are 1.5 to 2.1 times more likely than, respectively, domestic and local visitors to take and upload photographs of historical cultural heritage sites, which is consistent with the observations regarding their spatial distribution. By contrast, local inhabitants are 2.2 and 2.5 times more likely than international visitors to be associated with photographs reflecting aesthetic appreciation/mental health and birdwatching.

Discussion and conclusions

The results of this study support the notion that monitoring and analysis of the social media activity of the visitors to sites of environmental and historical importance may lead to an improved understanding of the spatial patterns of visitation and differences in how cultural benefits are accrued to various segments of the population. Insofar as the Usumacinta floodplain is concerned, the cultural services enjoyed by international tourists tend to be more spatially concentrated around sites of international importance, major urban centers and major roads, as well as more limited in the types of services enjoyed, with a lower appreciation of the local fauna and beauty of nature than demonstrated by local residents. This has potentially important implications for the sustainable management of the local natural capital and cultural heritage sites, because it allows policy makers to identify areas where overlapping interests may result in conflicts (e.g., around the area of Palenque) as well as to prioritize and tailor conservation policies to the specific and spatially differentiated demands of different segments of the population.

This study suggests that the analysis of the text associated with the titles and tags of geotagged photographs uploaded to social media sites may be a useful alternative or complement to more common approaches based on the analysis of the visual content of images. In spite of the obvious limitation that not all photographs are associated with a meaningful description, the fact that titles and tags are voluntarily assigned at a later time may lead to insights into the users' mental processes, personal conceptualization and memory of the scene that are not possible through the sole analysis of the image (Dror and Harnad 2008; Dunkel 2015). Tags may describe elements that are not directly visible in the image or provide insights into the perceived relative importance of different visual elements. We propose that future studies on geotagged photographs from social media will focus on developing techniques to systematically and conjointly tap the information that can be derived from both types of analyses.

Consistently with previous studies, our findings support the notion that visits by international tourists are more spatially concentrated than those of residents (Garcia-Palomares et al. 2015; Munoz et al. 2019) and that domestic visitors may differ from international visitors in interest they show for wilderness-related aspects (e.g., wild animals) (Munoz et al. 2019). The relatively high percentage of culturally tagged photographs we identified for the Usumacinta floodplain is comparable with the results of previous similar studies (Angradi et al. 2018; Van Berkel et al. 2018). Further comparison with the study by Angradi et al. (2018) on ecosystem benefits in the Great Lakes provides some additional insights. Similarly to Angradi et al.'s (2018) findings concerning the St. Louis River, we also observed differences in the content of photographs by local, domestic and international visitors in terms of fauna, albeit in the opposite direction, i.e., with local users taking more photographs of birds than other visitors. We did not however observe differences in terms of photographs of flora. Unlike Angradi et al. (2018), we also found that international tourists posted more photographs depicting cultural services than local residents and domestic visitors, which appears to be consistent with the fact that the Mayan ruins at Palenque are a major international tourist attraction.

In evaluating the results of this study, one should consider that some subjective judgment is unavoidable in the development of the set of keywords and association with specific cultural services. This problem appears to be shared by the bottom-up approach presented in the study and top-down approaches based on a priori definitions of keywords (e.g., van Zanten et al. 2016). To limit such investigator biases, some previous studies have chosen to rely on classifications based on the presence or absence in the photographs of specific elements (e.g., tourism infrastructure, recreational equipment, elements of fauna or flora) rather than established classifications of ecosystem services (Oteros-Rozas et al. 2018; Van Berkel et al. 2018). Another limitation in the interpretation of the results concerning the local population lies in the fact that domestic and international visitors are presumably more likely to upload photographs through their social networks, because unique events and situations are more likely to be shared (Ghermandi and Sinclair 2019; Wood et al. 2013). This implies that natural environments in remote locations that are not visited by tourists may be less suited for social media–based analyses (Becken et al. 2017). Finally, one should emphasize that although social media analysis can in principle be applied to a wider range of cultural services, in this study the analysis is limited only to the subset of cultural benefits that require physical presence and interaction with the environmental setting (Haines-Young and Potschin 2018).

In conclusion, the present study supports the notion that the wealth of information that the users of online social networking services daily upload and make publicly available on their profile webpages represents a valuable source of information for an improved understanding of how cultural ecosystem services and benefits from cultural heritage tourism are accrued to different categories of beneficiaries and in their spatial complexity. This, in turn, may be integrated in the policy discussion and decision-making processes to yield much needed, better-informed strategies for mainstreaming the conservation and sustainable management of natural capital and cultural heritage.

References

Angradi, TR, Launspach, JJ, Debbout, R (2018) Determining preferences for ecosystem benefits in Great Lakes Areas of Concern from photographs posted to social media. *Journal of Great Lakes Research* 44(2): 340–351.

Ardoin, NM, Wheaton, M, Bowers, AW, Hunt, CA, Durham, WH (2015) Nature-based tourism's impact on environmental knowledge, attitudes, and behavior: A review and analysis of the literature and potential future research. *Journal of Sustainable Tourism* 23(6): 838–858.

Asmelash, AG, Kumar, S (2019) Assessing progress of tourism sustainability: Developing and validating sustainability indicators. *Tourism Management* 71: 67–83.

Balmford, A, Green, JMH, Anderson, M, Beresford, J, Huang, C, Naidoo, R, Walpole, M, Manica, A (2015) Walk on the wild side: Estimating the global magnitude of visits to protected areas. *PloS Biology* 13(2): e1002074.

Becken, S, Stantic, B, Chen, J, Alaei, AR, Connolly, RM (2017) Monitoring the environment and human sentiment on the Great Barrier Reef: Assessing the potential of collective sensing. *Journal of Environmental Management* 203: 87–97.

Bojic, I, Massaro, E, Belyi, A, Sobolevsky, S, Ratti, C (2015) Choosing the right home location definition method for the given dataset. In: Liu, TY, Scollon, C, Zhu, W (Eds.). *Social Informatics: Lecture Notes in Computer Science*, vol. 9471. Springer, Cham, Switzerland.

Carabias, J, Sarukhán, J, de la Maza, J, Galindo, C (2010) *Patrimonio natural de México. Cien casos de éxito.* Comisión Nacional para el Conocimiento y Uso de la Biodiversidad, México.

Casalegno, S, Inger, R, DeSilvey, C, Gaston, KJ (2013) Spatial covariance between aesthetic value and other ecosystem services. *PloS One* 8(6): e68437.

Clancy, M (2017) *Tourism Research Information Network correspondence.* TRINET, Hawaii, 5 October.

Cohen, J (1960) A coefficient of agreement for nominal scales. *Educational and Psychological Measurement* 20(1): 37–46.

Coldwell, W (2017) First Venice and Barcelona: Now anti-tourism marches spread across Europe. *The Guardian*, 10 August. Retrieved on 19 January 2018 from www.theguardian.com/travel/2017/aug/10/anti-tourism-marches-spread-across-europe-venice-barcelona.

Connors, JP, Lei, S, Kelly, M (2012) Citizen science in the age of neogeography: Utilizing volunteered geographic information for environmental monitoring. *Annals of the American Association of Geographers* 102(6): 1267–1289.

Conrad, CC, Hilchey, KG (2011) A review of citizen science and community-based environmental monitoring: Issues and opportunities. *Environmental Monitoring and Assessment* 176: 273–291.

Dallen, TJ (2011) *Cultural Heritage and Tourism: An Introduction*, vol. 4. Channel View Publications, Bristol, UK.

Da Rugna, J, Chareyron, G, Branchet, B (2012) Tourist behavior analysis through geo-tagged photographies: A method to identify the country of origin. In: Proc. Of 13th International Symposium on Computational Intelligence and Informatics (CINTI). IEEE, Budapest, Hungary.

Donaire, JA, Camprubi, R, Gali, N (2014) Tourist clusters from Flickr travel photography. *Tourism Management Perspectives* 11: 26–33.

Dror, IE, Harnad, SR (2008) *Cognition Distributed: How Cognitive Technology Extends Our Minds*. John Benjamins Pub. Co, Amsterdam.

Dunkel, A (2015) Visualizing the perceived environment using crowdsourced photo geo-data. *Landscape and Urban Planning* 142: 173–186.

Elwood, S, Goodchild, MF, Sui, DZ (2012) Researching volunteered geographic information: Spatial data, geographic research, and new social practice. *Annals of the American Association of Geographers* 102(3): 571–590.

Figueroa-Alfaro, RW, Tang, Z (2017) Evaluating the aesthetic value of cultural eco-system services by mapping geo-tagged photographs from social media data on Panoramio and Flickr. *Journal of Environmental Planning and Management* 60(2): 266–281.

Garcia-Palomares, JC, Gutierrez, J, Minguez, C (2015) Identification of tourist hot spots based on social networks: A comparative analysis of European metropolises using photo-sharing services and GIS. *Applied Geography* 63: 408–417.

Ghermandi, A (2016) Analysis of intensity and spatial patterns of public use in natural treatment systems using geotagged photos from social media. *Water Research* 105: 297–304.

Ghermandi, A (2018) Integrating social media data analysis and revealed preference methods to value the recreation services of ecologically engineered wetlands. *Ecosystem Services* 31: 351–357.

Ghermandi, A, Sinclair, M (2019) Passive crowdsourcing of social media in environmental research: A systematic map. *Global Environmental Change* 55: 36–47.

Goodchild, MF (2009) Neogeography and the nature of geographic expertise. *Journal of Location Based Services* 3(2): 82–96.

Haines-Young, R, Potschin, M (2018) Common International Classification of Ecosystem Services (CICES) v5.1. Retrieved from www.cices.eu.

Hausmann, A, Toivonen, T, Heikinheimo, V, Tenkanen, H, Slotow, R, Di Minin, E (2017) Social media reveal that charismatic species are not the main attractor of ecotourists to sub-Saharan protected areas. *Scientific Reports* 7: 763.

Heagney, EC, Rose, JM, Ardeshiri, A, Kovac, M (2017) Optimising recreation services from protected areas: Understanding the role of natural values, built infrastructure and contextual factors. *Ecosystem Services* 31, Part C: 358–370.

Heikinheimo, V, Di Minin, E, Tenkanen, H, Hausmann, A, Erkkonen, J, Toivonen, T (2017) User-generated geographic information for visitor monitoring in a national park: A comparison of social media data and visitor survey. *ISPRS International Journal of Geo-Information* 6(3): 85.

Higham, J (2007) *Critical Issues in Eco-Tourism: Understanding a Complex Tourist Phenomenon*. Elsevier, Oxford.

Higham, J, Kearsley, G, Kliskey, A (2001) Multiple wilderness recreation management: Sustaining wilderness values, maximizing wilderness experiences. *The State of Wilderness in New Zealand* 81–93.

Hudson, PF, Hendrickson, DA, Benke, AC, Varela-Romero, A, Rodiles-Hernández, R, Minckley, WL (2005) Rivers of Mexico. In: Arthur Benke and Colbert Cushing *Rivers of North America* 2005. Elsevier Academic Press, Burlington, MA, pp. 1030–1084.

Koerbitz, W, Oender, I, Hubmann-Haidvogel, AC (2013) Identifying tourist dispersion in Austria by digital footprints. In: Cantoni, L, Xiang, Z (Eds.). *Information and Communication Technologies in Tourist 2013.* Springer Verlag, Berlin-Heidelberg, pp. 495–506.

Langemeyer, J, Calcagni, F, Baro, F (2018) Mapping the intangible: Using geolocated social media data to examine landscape aesthetics. *Land Use Policy* 77: 542–552.

Lee, H, Seo, B, Koellner, T, Lautenbach, S (2019) Mapping cultural ecosystem services 2.0: Potential and shortcomings from unlabeled crowd sourced images. *Ecological Indicators* 95: 505–515.

Lee, I, Cai, G, Lee, K (2014) Exploration of geo-tagged photos through data mining approaches. *Expert Systems with Applications* 41(2): 397–405.

Levin, N, Kark, S, Crandall, D (2015) Where have all the people gone? Enhancing global conservation using night lights and social media. *Ecological Applications* 25(8): 2153–2167.

Li, L, Goodchild, MF, Xu, B (2013) Spatial, temporal, and socioeconomic patterns in the use of Twitter and Flickr. *Cartography and Geographic Information Science* 40(2): 61–77.

Mancini, F, Coghill, GM, Lusseau, D (2018) Using social media to quantify spatial and temporal dynamics of nature-based recreational activities. *PLoS One* 13(7): e0200565.

Martinez-Pastur, G, Peri, PL, Lencinas, MV, Garcia-Llorente, M, Martin-Lopez, B (2016) Spatial patterns of cultural ecosystem services provision in Southern Patagonia. *Landscape Ecology* 31(2): 383–399.

Mihalic, T (2018) Anti tourism: A reaction to the failure or promotion for more sustainable and responsible tourism. Keynote. Travel and Tourism Research Association, European Chapter, Ljubljana.

Munoz, L, Hausner, V, Brown, G, Runge, C, Fauchald, P (2019) Identifying spatial overlap in the values of locals, domestic- and international tourists to protected areas. *Tourism Management* 71: 259–271.

Munro, J, Kobryn, H, Palmer, D, Bayley, S, Moore, SA (2017) Charting the coast: Spatial planning for tourism using public participation GIS. *Current Issues in Tourism* 1–19.

Oteros-Rozas, E, Martin-Lopez, B, Fagerholm, N, Bieling, C, Plieninger, T (2018) Using social media photos to explore the relation between cultural ecosystem services and landscape features across five European sites. *Ecological Indicators* 92(2): 74–86.

Richards, DR, Friess, DA (2015) A rapid indicator of cultural ecosystem service usage at a fine spatial scale: Content analysis of social media photographs. *Ecological Indicators* 53: 187–195.

Richards, DR, Tuncer, B (2018) Using image recognition to automate assessment of cultural ecosystem services from social media photographs. *Ecosystem Services* 31, Part C: 318–325.

Sanchez, AJ, Florido, R, Macossay-Cortez, A, Cruz-Ascencio, M, Montalvo-Urgel, H, Garrido-Mora, A (2012) Distribución de macroinvertebrados acuáticos y peces en cuatro hábitat en Pantanos de Centla, sur del Golfo de México. In: Sanchez, AJ, Chiappa-Carrara, X, Perez, B, *Recursos Acuaticos Costeros del Sureste*. CONCYTEY, Merida, pp. 416–443.

Sinclair, M, Ghermandi, A, Sheela, AM (2018) A crowdsourced valuation of recreational ecosystem services using social media data: An application to a tropical wetland in India. *Science of the Total Environment* 642: 356–365.

Spalding, M, Burke, L, Wood, SA, Ashpole, J, Hutchison, J, Ermgassene, PZ (2017) Mapping the global value and distribution of coral reef tourism. *Marine Policy* 82: 104–113.

Straumann, RK, Coltekin, A, Andrienko, G (2014) Towards (re) constructing narratives from georeferenced photographs through visual analytics. *The Cartographic Journal* 51(2): 152–165.

Stronza, A, Durham, W (2008) *Ecotourism and Conservation in the Americas*. CAB International, Cambridge, MA.

Su, S, Wan, C, Hu, Y, Cai, Z (2016) Characterizing geographical preferences of international tourists and the local influential factors in China using geo-tagged photos on social media. *Applied Geography* 73: 26–37.

Tenerelli, P, Pueffel, C, Luque, S (2017) Spatial assessment of aesthetic services in a complex mountain region: Combining visual landscape properties with crowdsourced geographic information. *Landscape Ecology* 32: 1097–1115.

Tenkanen, H, Di Minin, E, Heikinheimo, V, Hausmann, A, Herbst, M, Kajala, L, Toivonen, T (2017) Instagram, Flickr or Twitter: Assessing the usability of social media data for visitor monitoring in protected areas. *Scientific Reports* 7, 17615.

Tieskens, KF, Van Zanten, BT, Schulp, CJE, Verburg, PH (2018) Aesthetic appreciation of the cultural landscape through social media: An analysis of revealed preference in the Dutch river landscape. *Landscape and Urban Planning* 177: 128–137.

Tipaldo, G, Allamano, P (2017) Citizen science and community-based rain monitoring initiatives: An interdisciplinary approach across sociology and water science. *WIREs Water* 4: e1200.

Torre, A, Scarborough, H (2017) Reconsidering the estimation of the economic impact of cultural tourism. *Tourism Management* 59: 621–629.

Van Berkel, DB, Tabrizian, P, Dorning, MA, Smart, L, Newcomb, D, Mehaffey, M, Neale, A, Meentemeyer, RK (2018) Quantifying the visual-sensory landscape qualities that contribute to cultural ecosystem services using social media and LiDAR. *Ecosystem Services* 31(C): 326–335.

van Zanten, BT, Van Berkel, DB, Meentemeyer, RK, Smith, JW, Tieskens, KF, Verburg, PH (2016) Continental-scale quantification of landscape values using social media data. *Proceedings of the National Academy of Sciences* 113(46): 12974–12979.

Velwaert, M, Caley, P (2016) Citizen surveillance for environmental monitoring: Combining the efforts of citizen science and crowdsourcing in a quantitative data framework. *SpringerPlus* 5: 1890.

Vu, HQ, Li, G, Law, R, Ye, BH (2015) Exploring the travel behaviors of inbound tourists to Hong Kong using geotagged photos. *Tourism Management* 46: 222–232.

Wood, WA, Guerry, AD, Silver, JM, Lacayo, M (2013) Using social media to quantify nature-based tourism and recreation. *Scientific Reports* 3: 2976.

Wray, K, Espiner, S, Perkins, HC (2010) Cultural clash: Interpreting established use and new tourism activities in protected natural areas. *Scandinavian Journal of Hospitality and Tourism* 10(3): 272–290.

Yáñez-Arancibia, A (1999) Terms of reference towards coastal management and sustainable development in Latin America: Introduction to special issue on progress and experiences. *Ocean and Coastal Management* 42(2–4): 77–104.

Yañez-Arancibia, A, Day, JW (2004) Environmental sub-regions in the Gulf of Mexico: Eco-system approach as an integrated management tool in integrated coastal management in the Gulf of Mexico large marine ecosystem. *Ocean and Coastal Management* 47: 11–12.

Yáñez-Arancibia, A, Day, JW, Currie-Alder, B (2009) Functioning of the Grijalva-Usumacinta river delta, Mexico: Challenges for coastal management. *Ocean Yearbook* 23: 473–501.

6   Building consensus through assessment evidence from San Pedro de Atacama, Chile

Bernardo Broitman, Eric Sproles,
Craig Weideman, Sonia Salas,
Cristian Geldes, Antonia Zambra,
Leticia González-Silvestre and
Lorena Bugueño

Introduction

The work carried out in ProEcoServ-Chile (ProEcoServ-CL) was a collaborative effort led by the Centro de Estudos Avanzados en Zonas Áridas (CEAZA) in partnership with the Chilean Ministry of Environment and the municipality of San Pedro de Atacama (SPA), which is the pilot study area. During the last two decades, the territory of SPA has experienced significant socio-economic changes, mainly associated with the arrival of new social actors, including entrepreneurs, tourism operators and industrial mining. Economic development accompanying this new social diversity has, among other things, led to a marked increase in the local population size (estimated 50% increase from 2002 to 2014). Data from the National Institute for Statistics (Instituto Nacional de Estatistica, or INE) indicates that tourism in SPA has nearly tripled in the last decade, receiving an estimated 240,000 visitors in 2013 (INE 2014). The Los Flamencos National Park, located within the municipal boundaries, registered the second highest number of tourists of all Chile's protected areas in both 2013 and 2014, exceeding Torres del Paine in southern Patagonia. The massive influxes of tourism into the region in recent years has implied significant economic development potential for the municipality of SPA – each tourist budgets about US$1,000 to visit the destination. At the same time, however, these influxes imply a growing threat to the region's fragile high-altitude ecosystems and limited natural resources, particularly water, that sustain the livelihoods of its indigenous communities.

In light of this information, the main objective of ProEcoServ-CL was to develop innovative computer-based tools to guide decision making regarding sustainable management of two key ecosystem services (ESs): water and ecotourism (we apply the term to include both ecotourism and recreation activities). In particular, information on water provisioning and tourism flows was compiled and collated to support future policy and decision making regarding these ESs in the municipality of SPA. Considering relevant policy questions, and potential future development scenarios based on participatory approaches, ProEcoServ-CL has made this body of knowledge available to decision makers through a decision support system (DSS). The DSS was developed and designed with the participation of local and regional communities, aiming towards substantial capacity-building and the installation of a powerful tool for mainstreaming ESs into decision-making. The conservation and management of fragile Andean ecosystems, and the traditional lifestyles that take place there, require a commitment both from policy and decision makers, as well as the municipality's inhabitants; research embedded in social processes was therefore recognized as key to achieving sustainable development and effective management of ESs in the region (Cowling et al. 2008; Daily et al. 2009).

This chapter aims to synthesize the ProEcoServ-CL team's work in Chile, which included modelling water provision and ecotourism, developing a tool to support decisions and policies, and planning future scenarios considering the legal and economic context.

Objectives of the work

ProEcoServ-CL aimed to demonstrate the profound importance of incorporating ESs in the decision-making process in Chile. The main strategy was improved through the scientific and community-based understanding of social and ecological resources in the Salar de Atacama. Therefore, it combined quantitative and qualitative approaches to develop a baseline understanding of ESs in the region, identify the trends in social-ecological dynamics, and work within the community to develop the capacity to better manage ESs in the present and future. Furthermore, the physical presence of the ProEcoServ-CL team in the SPA community was crucial to the outstanding work of the local team. Members of the team assumed a range of responsibilities and tasks to complete the various project components, which was fundamental to the increased awareness and capacities now installed across a broad swathe of community members in relation to the targeted ESs. In this setting, the ProEcoServ-CL working plan was characterized by four main activities; the first three comprised: (1) modelling water provision and ecotourism (water balance model and ecotourism potential model); (2) developing a decision support tool (Tableau); and (3) building a participatory, inclusive decision-making process (with public and private stakeholders). Finally, ProEcoServ-CL (4) implemented various strategies for mainstreaming ecosystem services that delivered communication and outreach, co-production of knowledge, and policy intake. These strategies supported the integration of water provision services into spatial planning and dialogue, including the production of ES maps, the promotion of public-private cooperation for ecosystem management, and the establishment of a DSS framework that, in turn, has been used in spatial planning.

Modelling water provision and ecotourism

Assembling data

Data collection for the selected ESs, water provision and ecotourism, was accomplished with publicly available information for both, mostly from reports, which was confirmed or complemented through field observations and interviews. The Region of Antofagasta has a weak real-time monitoring system for the variables required for the analyses, which made it challenging to understand ESs dynamics in SPA, and institutions are not well prepared for contingent decision making.

In the case of ecotourism, quantitative data on tourism dynamics was identified as a key requirement for developing effective decision support capabilities. However, there was no system in place to monitor visitor numbers at tourist sites on a project-wide basis, largely due to the fact that there has been no coordination between local and regional institutions in this regard. At the time of writing, data are only available for a limited number of tourist sites in the study area, and, with the exception of those located within formally protected areas, are not efficiently managed or easily accessible in a centralized database. Therefore a key challenge was to explore novel approaches for near-real time monitoring of tourism activity across the project area. To this end, the team tested the use of geo-located tourist photos uploaded to public photo sharing websites such as Flickr. Flickr data have been shown to be positively correlated with actual visitation at more than 800 tourist sites globally (Wood et al. 2013), and were used in this study to derive information on spatial dynamics and relative levels of tourism activity across the project area as an indicator of ES use. The approach revealed several important insights relevant to decision support; for example, proportionally, just 10 sites account for a significant percentage of total tourism flows (30% of total annual visitor days), with 70% of annual visitor days distributed among the remaining ±80 recognized tourist sites. These activities also reinvigorated an initiative for the collection of tourism statistics in the Municipality of San Pedro de Atacama from 2014 to 2016, which was proposed in 2013 by the Fundación de Cultura y Turismo SPA. The initiative is currently being evaluated by INE, with the key objective of capturing tourism data from tour operators and hotels. The pilot proposal includes a 36-month period to identify trends and usage patterns in San Pedro de Atacama. For data collection, the project proposes the use of a monthly online survey that must be systematized by the regional offices (INE Antofagasta).

Improving systems for monitoring visitor numbers at the site level, understanding the social and ecological impacts, and developing appropriate management responses, will be fundamental to sustainable management of ecotourism in the region. In this regard, recommendations regarding sustainable visitor carrying capacities for several areas of ecotourism importance in SPA were sourced from an earlier study EUROCHILE (2006), and proved invaluable in developing decision support capabilities in our capabilities in our work.

In relation to water provisioning, ProEcoServ-CL found minimal reliable data for the Salar de Atacama, as the hydrological and meteorological systems in the Region of Antofagasta are sparse, and measured data commonly has a latency of several years. For example, streamflow data for the San Pedro River at Cuchabrachi (closest and most relevant stream gauge for the town of San Pedro de Atacama) was current through 2013. Precipitation and temperature data needed to estimate evapotranspiration was even more limited, as the meteorological station at El Tatio was current through 2002. Therefore, an explicit groundwater model for the region was not feasible due to the extremely limited data. The project, therefore, focused on the San Pedro Watershed, an important watershed for the region that has sufficient, if limited, data from the Dirección General de Aguas (DGA) website, Dirección General de Aguas (2015), which could be used for the basic statistical analysis of water resources trends and the development of a conceptual water balance model of the storage and fluxes of water in this hydrologically closed basin. The ProEcoServ-CL team's approach to examining water provision throughout the Altiplano/plateau region was suggested by the local Steering Committee, and was based on remote sensing using NASA's Gravity Recovery and Climate Experiment (GRACE) instrument (Tapley et al. 2005). This satellite-based system measures changes in the Earth's mass through time and can reliably detect changes in groundwater. Analysis of GRACE data for the region of the Altiplano where San Pedro de Atacama lies proved a key milestone for ProEcoServ-CL. This data produced the first hydrological balance for the region (Figure 6.1).

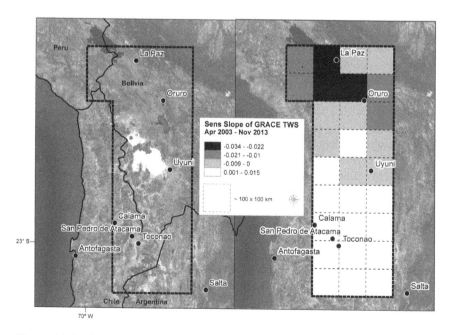

Figure 6.1 Satellite hydrological balance.

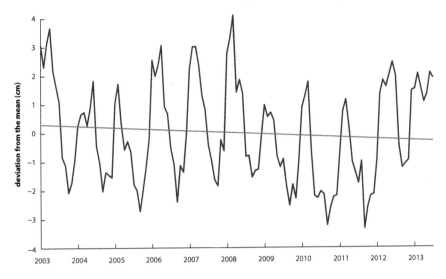

Figure 6.1 (Continued)

Table 6.1 Statistical values for annual Precipitation
(P) and streamflow (Q) for 1978 to 2010

	P	Q
Coef Var (mm)	0.75	0.32
Median (mm)	123.6	20.0
STD (mm)	104	6.4
P-value	0.59	0.68

Source: Author's computation

The analyses showed that during the mission period examined (2003 to 2013) there was basically no trend in groundwater levels (2% slope over the period). This single result, together with a comparison to literature showing the situation in the Middle East (200% slope) (Longuevergne et al. 2013), emphasized the need to move from discussions governed by perceptions to activities that capture information locally and in a reliable fashion.

Analysis of precipitation (P) and streamflow (Q) data shows variability with regards to precipitation, but minimal variability with regards to streamflow. From 1978 to 2013, annual precipitation represented a high degree of variability, and annual streamflow remained relatively constant across all years (Table 6.1). The coefficient of variation, the standard deviation divided by the mean (i.e. a normalized measure of variation within a data set), for annual P (0.75 mm) was roughly 2.5 times greater than annual Q (0.32 mm). These data demonstrate that even in years of high or low precipitation, streamflow remains fairly consistent across years, strongly suggesting that groundwater is a major contributor to streamflow.

126   *Bernardo Broitman et al.*

Furthermore, there were no statistical trends associated with these data, indicating that, while variability occurs across years, precipitation and streamflow have not shown any considerable changes during the study period.

Water balance model

The conceptual water balance model was run for 1992 to 2002, based upon the data available from the DGA. The results support the conclusions from straight data analysis – groundwater fluctuations dominate the hydrological cycle even on a monthly basis (Figure 6.2). From 1993 to 1995, nine individual months with more than 12 mm of precipitation occurred. These inputs are not expressed in streamflow (Q), but are likely responsible for considerable increases in groundwater storage (S). The following year, 1996, has markedly less precipitation. However, Q remains consistent, but S decreases. This suggests that the nine precipitation events occurring from 1993 to1995 increased groundwater stores, which then sustained base flows during the subsequent, drier year. The years from 1997 to 2000 each have one month with precipitation over 25 mm, and groundwater recharge fluctuates on an annual cycle.

This same modelling framework was applied to 40-year mean temperature and precipitation data for the region using a distributed data set for the Rio San Pedro

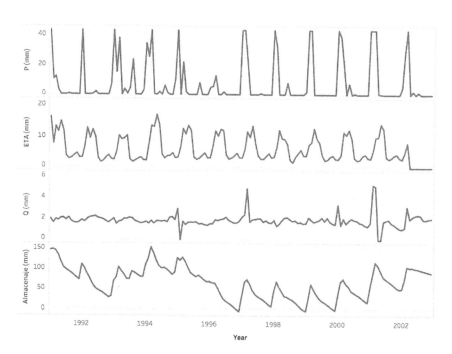

Figure 6.2 Model results for the Rió San Pedro Watershed based on hydrometeorological data from the Dirección de General de Aguas (www.dga.cl/servicioshidro meteorologicos/Paginas/default.aspx).

and adjacent Rio Vilama. This approach provides an overview of groundwater storage and the timing and magnitudes of water fluxes, highlighting December as the month of greatest water scarcity.

We next tested the sensitivity to increases or decreases in precipitation and temperature, perturbing inputs by ±10% in order to simulate climate variability. Each perturbation was run individually for five total model iterations for 12 months. The results show that December is the most climatologically sensitive month with regards to groundwater, and that variability can range by 25%. From a management standpoint this is important, as December corresponds with greatest water scarcity and the onset of the peak tourist season and irrigation. From June to September, the model suggests there is minimal sensitivity to climate variability. The water resources during this time are minimal, but consistent, due in large part to groundwater storage during the wet season (January to March). These results demonstrate that climate variability is not expressed equally across the region. The results provide a conceptual analysis and numerical values should be interpreted with caution. This highlights the lack of capacity for a current interpretation of water resources dynamics in SPA, and its institutions are not well prepared for contingent decision making.

Ecotourism model

Recognizing ecotourism as both a service of ecosystems as well as an important driver of ecosystem change (Millennium Ecosystem Assessment, 2005), developing decision support capabilities for management focused on generating the information required to characterize key variables and feedbacks in linked social-ecological systems in the project area. ProEcoServ-CL integrated an array of qualitative and quantitative approaches to evaluate these interactions, including local knowledge leveraged through participatory processes to inform multi-criteria decision analyses, public data from web-based social media platforms, GIS analyses, and results from a range of earlier studies, providing a basic framework for users to proactively interrogate management interventions according to specific decision-making contexts.

Quantifying ecotourism benefits

Cultural ESs manifest primarily as an expression of subjective individual or societal values attached by humans to nature, and, as such, are only indirectly linked to ecosystems themselves (Hernandez-Morcillo et al. 2013). Efforts to quantify cultural ESs, including ecotourism, have focused on the use of social processes to interpret these values, explicitly linking them to ecosystem structures and processes and other relevant site-level attributes to produce spatially resolved estimates of ES provision across the landscape (Nahuelhual et al. 2013; Casado-Arzuaga et al. 2013; Peña et al. 2015). To map ecotourism benefits in SPA, we convened a specialist focus group comprising local industry professionals to leverage collective knowledge of how tourists interact with the environment, applying formal decision

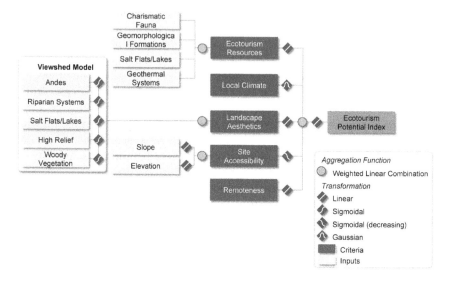

Figure 6.3 Decision hierarchy showing spatial attributes and aggregation and transformation functions applied at each step to derive an index of ecotourism potential (EPI).

rules and procedures based on the Analytic Hierarchy Process (AHP) and pairwise comparison (Saaty 2008) to evaluate a suite of environmental attributes relevant to ecotourism (Figure 6.3). Spatial layers representing each attribute, or criterion, and inputs were created using standard GIS analyses and synthesized according to respective group priority weights to produce a map of ecotourism potential (or Ecotourism Potential Index) reflective of relative levels of service provision across the landscape (Figure 6.4).

To aid visual interpretation and downstream analyses, index values were classified according to natural breaks to produce five ecotourism potential (EP) classes ranging from low (1), medium-low (2), medium (3), medium-high (4) to high (5) EP. By this approach, 2.3 % of the total project area was classified as high EP (5), with nearly half, or 48.9%, qualifying as low EP (1); classes 2, 3 and 4 accounted for 28.6%, 16.1% and 4.1% of the project area respectively (Figure 6.5). Highest EP values were generally found to occur in association with salt flats and lakes located on the Altiplano east of the cordillera, coinciding with areas of high aesthetic value and animal diversity, with areas of lower potential occurring predominantly in the south and west of the Salar de Atacama.

Mapping visitor trends and defining sustainable use limits

Site-level visitor data offer key insights for researchers and managers, both in terms of understanding how people interact with ecosystems when engaging in

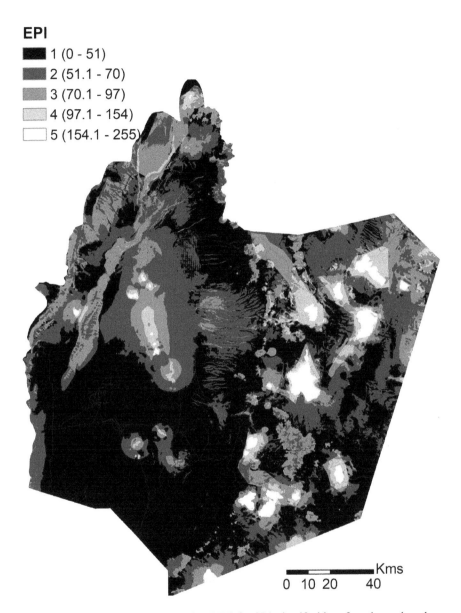

EPI
■ 1 (0 - 51)
■ 2 (51.1 - 70)
■ 3 (70.1 - 97)
□ 4 (97.1 - 154)
□ 5 (154.1 - 255)

Figure 6.4 Ecotourism Potential Index (EPI) for SPA classified into five classes based on Jenks natural break optimization, from 1 (low) to 5 (high EP).

ecotourism activities, as well as monitoring potential environmental impacts associated with varying levels of tourism use across the landscape. Developing specific
science-based recommendations for mitigating tourism impacts on ecosystems
is challenging, but, despite valid criticisms, the concept of a tourism carrying

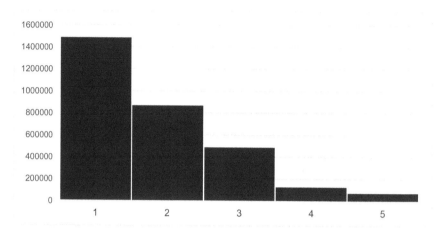

Figure 6.5 Histogram of frequency distributions of the five EP classes in Figure 6.4.

capacity (TCC), typically defined on the basis of inferred site-level physical, eco-logical and sociopsychological thresholds, has frequently been used to establish clear numerical targets for managers (McCool et al. 2007). TCC recommendations were developed for several established ecotourism sites in SPA in an earlier study (EUROCHILE 2006) and were integrated into our approach to improve decision support capabilities in this context (Figure 6.6).

Figure 6.6 shows visitor numbers plotted against TCC at El Tatio geysers (site "a" in Figure 6.7), one of the sites for which both visitation and TCC data overlap in 2012 and 2013, indicating that visitation exceeded recommended TCC on 81% and 65% of days for which data were available in respective years.

The utility of the approach in this case is currently limited by the availability of empirical site-level visitor data, discussed earlier, and is a widely reported prob-lem in the ecotourism literature (Wood et al. 2013). While plans are in place to address this gap in SPA, a proxy method for predicting visitation was tested as a possible alternative, based on geotagged tourist photographs uploaded to the pub-lic web-based photo sharing platform Flickr (Millennium Ecosystem Assessment 2005) (Figure 6.7). Although regression analysis revealed a poor fit of photo and empirical data, potentially due in part to the relatively small number of data points available at the spatial scale of this analysis, results could be improved by more careful filtering of inaccurately geotagged photos from the dataset.

With this caveat, the approach was nevertheless assumed to be representative of broad trends in tourism patterns, and revealed several insights relevant to manage-ment; for example, just 10 sites were found to account for a significant proportion of total Flickr "photo user days" (33% of total annual Flickr user days), with 67% of user days distributed among the remaining ±80 recognized tourist sites. However, although Flickr-based maps of tourist activity broadly correspond with the EPI developed on the basis of professional judgement (Figure 6.4), results indicate poor spatial agreement between areas of high predicted EP and Flickr use

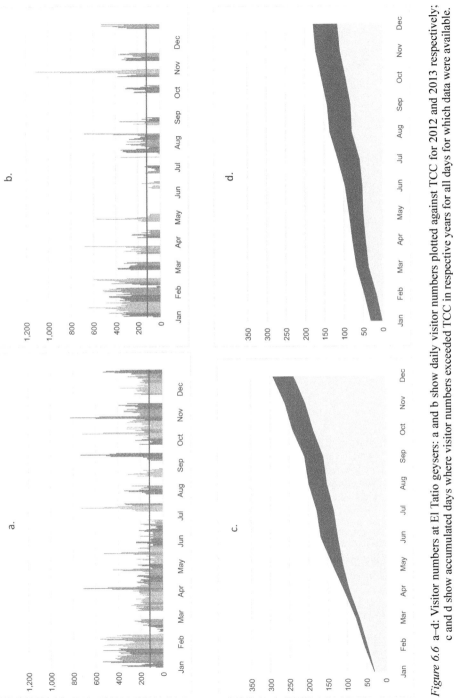

Figure 6.6 a–d: Visitor numbers at El Tatio geysers: a and b show daily visitor numbers plotted against TCC for 2012 and 2013 respectively; c and d show accumulated days where visitor numbers exceeded TCC in respective years for all days for which data were available.

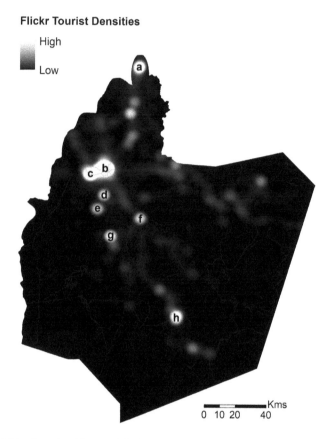

Figure 6.7 Tourism densities modelled according to unique geo-located Flickr user tags
recorded at each 90m grid cell in the study area; locations of major tourist sites
are labelled a–h, including: El Tatio geysers (a); San Pedro de Atacama town (b);
Valle de la Luna (c); Lake Cejar (d); Lake Tebenquiche (e); Jere/Toconao (f);
Lake Chaxa (g); and lakes Miscanti and Miniques (h).

particularly in the east of the project area; this may reflect a need to refine the EPI,
or alternatively, that current patterns of use are in fact sub-optimal with respect to
the supply of ESs across the landscape.

Evaluating impacts of key drivers of change

In addition to ecotourism, mining was identified as among the most important
local drivers of change in SPA. Quantifying and characterizing the range of direct
and indirect impacts of these processes on ecosystems and ESs is complex (Millennium Ecosystem Assessment 2005), and was not attempted for the purposes
of decision support in this study. However, degradation of aesthetic quality of the
environment, highly valued in the context of ecotourism both in this study and in

the literature (Casado-Arzuaga et al. 2013), associated with the expansion of built infrastructure is an immediately recognizable direct impact linked to mining and development on ecotourism in SPA. Visual impacts can be relatively easily modelled in a GIS environment based on line-of-sight calculations and freely available digital elevation data. Interrogating the results of visual impact analyses against EPI classes enables decision makers to proactively evaluate development options that objectively imply least impact on areas of high ecotourism potential.

As proof of concept, an analysis of visual impact associated with mining infrastructure located in the south of the project area, intersected with the EPI (Figure 6.4), revealed negligible visual impacts in areas mapped as high and medium-high ecotourism potential (Figures 6.8 a–b); EP classes 5 and 4 (low and medium-low) were determined to experience the most visual impact in terms of area from mining, although impact severity is predominantly low (Figure 6.9).

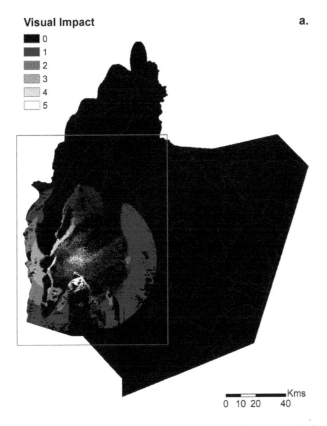

Figure 6.8 (a) Visual impact of mining infrastructure located in the south of the Salar de Atacama: impact is expressed in terms of the proportion of infrastructure visible from each pixel (subject to atmospheric conditions) classified into 5 impact severity classes based on natural breaks in values, where 1 = low impact and 5 = high impact; (b) visual impact intersected with the five EP classes to derive visual impacts per EP class; the legend reports EP class followed by visual impact severity.

b.

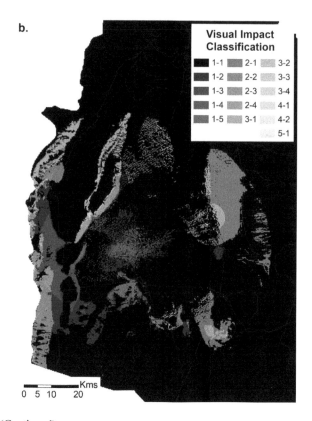

Figure 6.8 (Continued)

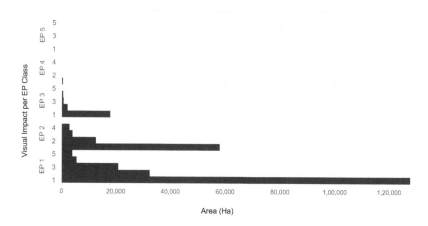

Figure 6.9 Visual impact severity per EP class.

Source: Author's own

Economic valuation pilot exercise

Considering the data assembled, as well as the water and ecotourism models, the ProEcoServ-CL team performed an economic valuation exercise. The valuation scenario was discussed with local and regional stakeholders, taking into account the fact that a changing scenario would affect them more directly than it would national actors. This work extended over several workshops. In this framework, 700 survey questionnaires were distributed to local stakeholders and main actors involved in productive and development activities in the region, including mining companies, tourism operators, NGOs, and small and medium-sized enterprises (SMEs). The Unit of Environmental Economics from the Ministry of the Environment was involved in the design of the valuation scenarios. The underlying strategy for economic valuation of the ES of interest, water provision, was to work with several actors involved in the local, regional and national economy to influence economic development planning (see economic value estimates for the tourists, Box 6.1).

Box 6.1   Contingent valuation study in San Pedro de Atacama

ProEcoServ-CL administered an exploratory economic valuation study with the use of questionnaires, which is known in the literature as contingent valuation. In this valuation exercise, tourists are asked to express their preferences with respect to a water management scheme that is characterized by guaranteeing a steady and continuous provision of water in relation to the environmental health, hydrological and living resources of the region of the Salar de Atacama. Estimation results show that tourists' maximum willingness to pay ranges between US$6.90 and $11.32. This corresponds to approximately 1% of the budget reported for visits to the area. Taking into account the visitor numbers for San Pedro de Atacama, about 260,000 according to the latest figures published by the INE, the introduction of a payment scheme that would collect such a monetary amount would create an additional annual revenue to the municipality in the range of US$1.79 to $2.94 million. This corresponds to a revenue of US$20 per inhabitant of the El Loa Province, where the Salar de Atacama is located.

Source: CEAZA – PROECOSERV Report 1143/2014 (UNEP 2015)

These results were discussed in several workshops, and demonstrated how SPA is valued from both internal and external perspectives; in other words, stakeholders were able to understand not only how they value their ecosystems, but also how external actors perceive the destination in economic and productive terms (e.g. attractive landscapes and cultural attributes for tourism). Based on the results, dissemination material and practical guidelines were produced to create widespread impact at the national level.

Developing a decision support tool

Working with Tableau software

One of the key goals guiding ProEcoServ-CL's work was the development and application of a tool that would allow streamlined access to information gathered by the initiative and validated by the local community. By designing and distributing the tool, and then training its users, ProEcoServ-CL's team sought to mainstream ESs, thereby empowering stakeholders to participate with policy and decision makers involved in spatial planning and management of ecosystems and the ESs they supply. The participatory process used to design the tool allowed local and regional actors to be part of the data collection, modelling, assessment, and mapping of ESs. The protracted process was key to strengthening the adaptive governance capacities and social learning skills regarding ecosystem services.

On the other hand, this adaptive participatory process hinged on the identification, discussion and planning of scenarios for San Pedro de Atacama. During the four years of ProEcoServ-CL, several surveys, workshops and meetings were conducted to build future scenarios for the municipality, allowing a broad array of representatives from the national, regional, and local levels to participate in these collective exercises. The results revealed a range of perceptions regarding future scenarios over 10, 30, 50, and 100-year planning horizons. The capacity of the local population to respond to these conditions was synthesized in two main potential scenarios. The first scenario was based upon a "social management of water resources," which envisages that adverse conditions arising from increasing tourism will be compensated for by technological solutions and innovation, and will support better use and planning of ESs, new policies and the opening of new markets. The second scenario was based on "water resources and social mistrust," anticipating that water scarcity will precipitate a major crisis involving biodiversity loss, human migration and diseases. This crisis will generate conflicts of interest, increasing social mistrust. To allow user-friendly access to these results, ProEcoServ-CL's team developed a DSS tool based on Tableau 8.2 software, which is available to all stakeholders from the community and institutions. The same stakeholders are committed to maintain this DSS and provide access to anyone interested in the data. This software is a data analysis platform that is easy to learn and use (Figure 6.10).

Using multiple alternatives for data connection (Excel, servers, etc.), users can visualize trends and dynamics associated with modelling work of ESs. Furthermore, based on the scenario results mentioned previously, users can generate and share scenarios using this data. The software allowed government institutions and organizations, at regional and local levels, to participate in the design and training of the DSS. Government institutions represented at the regional level included the Regional Undersecretariat (SEREMI) of Medio Ambiente (Ministry of the Environment, Antofagasta) and SEREMI de Agricultura (Ministry of Agriculture, Antofagasta). Institutions represented at both the regional and local level included the Corporación Nacional Forestal (National Forest Service, CONAF) and Servicio Nacional de Turismo (National Tourism Service, SERNATUR). Finally, the Consejo de Pueblos Atacameños (Atacameño Peoples' Council), Ilustre Municipalidad de San Pedro de Atacama, Fundación de Cultura y Turismo, Asociación

Figure 6.10 Training workshops in Tableau software, San Pedro de Atacama, January 2015.

de Regantes Rio San Pedro, Fundación Tata-Malku, and Asociación de Turismo y Medio Ambiente (ATYMA) were represented at the local level.

Assessment and evaluation

The overall effectiveness of the assessment and evaluation of these ESs analyses and models was monitored throughout the whole process and its design. Additionally, hands-on workshops provided an opportunity for business leaders, decision makers and students from the municipality to apply Tableau software beyond simple "button pushing" of the DSS. The workshops included 10 participants, and placed considerable focus on how to continue development and application of the tool for future decision making. Participants were given the opportunity to ask and answer questions, review the database that was being used, and experiment with different types of data visualization using the software.

Workshop effectiveness was evaluated through surveys. Participants took surveys before and after the workshops to assess their self-efficacy and abilities in applying data and technology in decision making and problem solving (Hiebert 2012). After the training workshops, participants showed an overall 24% increase in self-efficacy. Participants with less experience in natural resources or engineering (primarily business leaders) displayed even greater improvements in self-efficacy (>50%). Additionally, the performance of the workshops was evaluated by asking participants what they liked and did not like about the workshops, and what improvements should be made. These assessments allowed the ProEcoServ-CL team to address students' learning needs and maximize the impact of the DSS.

Public discussion, involvement and participatory process

When designing spatial mapping products for ESs, one of our major goals was to implement them locally to increase participation and trust between stakeholder and policy makers, ensuring the long-term continuity of ProEcoServ-CL results. Using Tableau, the ProEcoServ-CL team developed dashboards, including the model results for both ESs, to be applied in a decision support context. Furthermore, during the closing activity (March 19, 2015), the team presented the DSS tool tutorial, hosted on the ProEcoServ-CL webpage. We highlighted some of the main results and showed some of the activities that ProEcoServ-CL developed with stakeholders during the training workshops in January 2015.

Additionally, the team produced and delivered workshops, activities, meetings, documents and material during this final period to increase awareness and capacity-building regarding ESs at different levels. For example, ProEcoServ-CL coordinated and implemented an educational program at San Pedro and Talabre schools, named "Los caminos de la Patta Hoiri," with the main objective of raising awareness among primary and secondary students about the ecosystems and natural environment around San Pedro de Atacama. Furthermore, in December 2014 we hosted the final dissemination workshops for the Atacameño indigenous communities. Workshops were conducted in seven different communities including Solor, Rio Grande, Yaye, Socaire, Talabre, Toconao and Sequitor.

Social learning, communication and adaptive governance

Overall, the strategy of the ProEcoServ-CL project was to strengthen the awareness and understanding of ESs at the local level, with a particular emphasis on the social learning process behind this strategy. This is widely recognized as one of the fundamental requirements for successfully operationalizing the ES concept (Cowling et al. 2008; Daily et al. 2009; Sitas et al. 2013). This approach promoted better communication with regional and national policy and decision makers. It is critical that national policy makers better understand how policy decisions affect populations whose livelihoods depend directly on different ESs (e.g. you cannot run out of water in the desert). For this reason, ProEcoServ-CL coordinated a series of workshops. One set of workshops was designed to disseminate a strategy for SMEs and entrepreneurs at the local level including NGOs, guilds, organizations, communities and companies linked to ecotourism in the municipality. This activity was aimed at establishing a strategic plan, and underlying policy instruments, to value and develop ecosystem services in the context of SMEs administration. For example, within the indigenous communities and ecotourism operators, there are concerns about minimizing the loss of cultural identity and biodiversity, while managing sustainable economic growth, especially with regards to changes in water availability. From a scientific perspective, interpreting changes in the flow of ESs and the impacts was constrained by a sparse monitoring network and a limited understanding of ESs processes in the region. At a human level, concerns regarding changes in water resources and tourism impacts are unique across individual communities. This dynamic requires appropriate research initiatives to balance scientific objectives and community needs within the realities of limited data. From a governance perspective, the assessment survey used in the DSS training workshops also provided information about capacity levels among stakeholders and decision makers. This information is useful, as it will allow improved training and dissemination of ProEcoServ-CL project results to other members of the community, thus increasing awareness and capacity-building in relation to the feedback from the surveys. Furthermore, communication and coordination between policy makers was key to promoting better planning and management of ESs in the country. Chile has already taken on the task of implementing ESs in the national policy. Outcomes from ProEcoServ-CL, delivered though activities, workshops, dissemination material, educational programs and a community-based DSS are tangible guidelines for future national level initiatives designed to sustain ESs.

To conclude, scientific and community-based understanding of ESs has been demonstrated as a key ingredient throughout the participatory process, and it underpins the outcomes of ProEcoServ-CL's work. During this process, the presence of a local team was a key factor for consolidating and strengthening the exchange between science and political knowledge. Furthermore, many key actors, including citizen organizations, government institutions and technical experts, collaborated on the objective in many ways. For example, these actors provided information, systematized data, provided physical spaces, disseminated information and participated in the activities that ProEcoServ-CL organized.

Implementation, outreach and enforcement

The ProEcoServ-CL team aimed to prepare the institutional and organizational conditions required to maintain and monitor the DSS for the ESs package. These conditions implied strong organizational connections between decision makers, policy makers and stakeholders. These connections were crucial to the development of the ESs decision tool. Two workshops at the local and regional level (SPA and Antofagasta) were developed to accomplish this objective. One of the strategies during these meetings (to assure long-term continuity of ProEcoServ-CL's results) was to maintain a copy of the database inside CEAZA's computational infrastructure while decision makers and stakeholders decide which institution will be responsible for updating and maintaining the DSS. The systematization and validated data provision will depend on the collaboration of several institutions related to both ESs: Corporación Nacional Forestal (CONAF), Servicio Nacional de Turismo (SERNATUR), Seremi Medio Ambiente (SEREMI MA) and Dirección General de Aguas (DGA). An interesting outcome of this particular objective was to learn that none of the different government institutions were particularly qualified or willing to take on the responsibility. This emergent finding reinforced one of our strategic recommendations, which was to implement the ESs strategy within locally constituted, intermediate decision-making bodies. This is likely one of the keys to the long-term sustainability of the ProEcoServe-CL results, and is currently being promoted in the national ESs agenda.

Training and capacity building

ProEcoServ-CL, along with decision-makers and stakeholders, decided to use Tableau software as the DSS for both water provision and ecotourism. This platform was ready in December 2014. At the end of January 2015, an intensive two-day workshop was organized in San Pedro de Atacama as the first pilot implementation, dissemination and training of the DSS to decision makers and users. This workshop was intended to empower locals to better understand the local-level processes affecting both ecosystem services and to provide them with the information required to better manage resources.

During the project's closing activity, the ProEcoServ-CL team also provided a book to local and regional communities called Memoria de Gestión ProEcoServ 2011–2015. This book summarizes the major work products of the ProEcoServ-CL project, and also seeks to help decision and policy makers understand and manage selected ESs at different levels of decision and policy making. This particular outreach product also provides some leads for the online training material for stakeholders that was launched along with a webpage (http://proecoserv.ceaza. cl/) that will permanently post all of the key deliverables, training material and databases (Figure 6.11).

Figure 6.11 Guideline and promotional material: *Memoria de Gestión ProEcoServ 2011–2015.*

Source: Centro de Estudios Avanzados en Zonas Áridas; Memoria – ProEcoServ Chile; published by Centro de Estudios Avanzados en Zonas Áridas; La Serena, Chile, 2015; http://proecoserv.ceaza.cl/pdf/memoria.pdf

Resultados de "ejercicios de escenarios" en la comuna San Pedro de Atacama.

Infografía elaborada a partir de los resultados obtenidos de las encuestas y talleres participativos con diversos actores locales y regionales. A continuación detalles de los escenarios futuros proyectados para la comuna de San Pedro de Atacama.

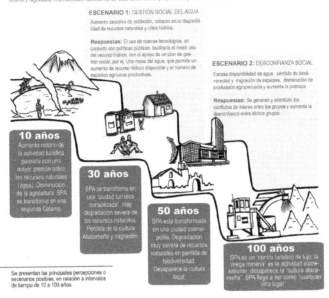

ESCENARIO 1: GESTIÓN SOCIAL DEL AGUA

Aumento excesivo de población, colapso en la disponibilidad de recursos naturales y crisis hídrica.

Respuestas: El uso de nuevas tecnologías, en conjunto con políticas públicas, facilitaría el mejor uso del recurso hídrico, con el apoyo de un plan de gestión social, por ej. Una masa del agua, que permita un aumento de recurso hídrico disponible y el número de especies agrícolas productivas.

ESCENARIO 2: DESCONFIANZA SOCIAL

Escasa disponibilidad de agua, pérdida de biodiversidad y migración de especies, disminución de producción agropecuaria y aumenta la pobreza.

Respuestas: Se generan y acentúan los conflictos de interés entre los grupos y aumenta la desconfianza entre dichos grupos.

10 años
Aumento notorio de la actividad turística pareada con una mayor presión sobre los recursos naturales (agua). Disminución de la agricultura. SPA se transforma en una segunda Calama.

30 años
SPA se transforma en una "ciudad turística consolidada". Hay degradación severa de los recursos naturales. Pérdida de la cultura Atacameña y migración.

50 años
SPA está transformada en una ciudad cosmopolita. Degradación muy severa de recursos naturales en pérdida de biodiversidad. Desaparece la cultura local.

100 años
SPA es un "centro turístico de lujo, la "mega minería" es la actividad sobresaliente: desaparece la "cultura atacameña". SPA llega a ser como "cualquier otro lugar".

Se presentan las principales percepciones o escenarios posibles, en relación a intervalos de tiempo de 10 a 100 años.

Lugar: Comunidad atacameña Río Grande

Figure 6.11 (Continued)

1 Presentación.

2 Marco conceptual: Los servicios ecosistémicos y el bienestar humano.

3 La comuna de San Pedro de Atacama: El sitio piloto del ProEcoServ en Chile.

4 Herramientas de apoyo a la participación ciudadana.

4.1   Modelación de agua en la subcuenca del Río San Pedro.
4.2   Modelación de potencial ecoturístico y su vínculo con actividades humanas.
4.3   La Plataforma Tableau: Una herramienta de integración para gestión de servicios Ecosistémicos.
4.4   Abriendo espacios para el encuentro de saberes: El programa educativo de ProEcoServ.

5 Antecedentes a considerar en las políticas para el desarrollo sustentable.

5.1   Escenarios futuros; Evaluando las condiciones ecosistémicas y el bienestar humano en la comuna de San Pedro de Atacama.
5.2   Construyendo un futuro sustentable en San Pedro de Atacama: Estrategia y valoración económica de servicios ecosistémicos.
5.3   Los servicios ecosistémicos en el marco de las políticas ambientales: Una propuesta para su inclusión a nivel regional y local.

6 Consideraciones finales.

7 Bibliografía.

Figure 6.11 (Continued)

Mainstreaming ecosystem services and influencing policy

Participatory decision making at the local level

One of the most important results from ESs mainstreaming during the project implementation was the social learning process attached to each one of the objectives. This approach helps strengthen the exchange of ESs knowledge between stakeholders at the local, regional and national levels, as well as a community-based understanding of ESs which, linked with a sectoral political framework, can also support planning at the national level. Therefore, one of the main proposals from ProEcoServ-CL, in association with a political strategy to mainstream ESs into sustainable national development planning, is the central role of participatory decision making at the local level. The proposal aimed to link policy support tools at different levels based on local and multi-sectoral interests that promote actions to protect and conserve ecosystems relevant for national development.

Considering this strategy, each outcome can be introduced into the policy support tools selected, especially at the local level, to influence national development planning. The ESs mainstreaming through a participatory process was already validated by many of the project outcomes: (1) identification and weighting, together with stakeholders, of spatial factors that determine areas of ecotourism potential in San Pedro de Atacama; (2) developing and training of models for water provision and ecotourism using Tableau software; (3) training and dissemination of trade-off matrices for both ESs; (4) identification and review of scenarios for San Pedro de Atacama, including the design of final guideline material on this issue and (5) conducting several meetings to identify potential SMEs and engaging with other private-public projects.

The ProEcoServ-CL team worked closely with 12 government institutions and organizations participating in a local steering committee (LSC) that was chaired by the Municipality of San Pedro de Atacama. During the period of project implementation, the group worked closely with the ProEcoServ-CL team to, above all, link activities and outcomes with national, regional and local policy processes, and also to provide important feedback for research. Each government institution chosen for the LSC was related directly, or indirectly, with the ESs targeted by the proposal. Within the government institutions, there was an advisory group that also supports several other participatory processes at the local level. The main institutions and organizations that were part of one or both groups were the Municipality of San Pedro de Atacama (Municipalidad de San Pedro de Atacama), the Council of Atacameños People (Consejo de Pueblos Atacameños, CPA), Rio San Pedro Irrigators and Farmers Association (Asociación de Regantes y Agricultores Río San Pedro), National Service of Tourism (Servicio Nacional de Turismo), National Forestry Corporation – Los Flamencos Reserve (Corporación Nacional Forestal), Water Authority (Dirección General de Aguas), Foundation of Culture and Tourism (Fundación de Cultura y Turismo), Regional Ministry of Agriculture, Secretary of Regional and Administrative Development (Subsecretaría de Desarrollo Regional y Administrativo), National Indigenous Corporation (Corporación Nacional Indígena), Association of Tourism and Environment (Asociación de Turismo y Medio Ambiente) and Tata Malku Foundation (Fundación Tata Malku). Additionally, at the local level, ProEcoServ-CL identified the Communal Development Plan and the Municipal Ordinance for Local Environmental Management as the two major entry points for policy intake. It is worth noting that ESs are not clearly integrated into this political and legal framework, although the ecosystem concept is mentioned in the higher-level policy tools.

Engagement with regional and national development planning

The core of the methodological proposal was a participatory decision-making process at the local level. In this sense, the major concern was to generate a strategy for the sustainable management of ESs from a multi-sectorial perspective, which articulated several actions to protect the ESs in San Pedro de Atacama and ensure the well-being of its people. Innovation and adaptive governance for a sustainable development of ESs in San Pedro de Atacama depended on the strategies used to link national policy tools with regional and local decision tools and decision making. One of the key lessons learned is that participatory processes, together with a scientific and community-based understanding of ESs, have been instrumental in reaching these goals, with the result that the community is now truly empowered to manage their ESs.

At the regional level, ProEcoServ-CL identified two main tools that included ESs in policy and decision making: Regional Development Strategy and the Action Plan for Biodiversity in the Antofagasta Region. The ProEcoServ-CL team also collaborated with the Ministry of Environment, including the regional undersecretary's office. Hence, it was possible to work very closely, enabling the minister to participate in important activities and discuss perspectives regarding ESs, especially in relation to the design and training of the DSS. Furthermore, the ProEcoServ-CL local team was able to link several national and regional initiatives. A concrete achievement in this regard was the submission of a proposal to Antofagasta's Regional Fund for Innovation and Competitiveness that aimed to strengthen the capacity of regional decision and policy makers to analyze decisions on key ESs. ProEcoServ-CL also secured funding for two projects in the 2015 period. The Environmental Protection Fund, sponsored by the Ministry of the Environment, is supporting a renewable energy project called Ckapin isaya Ckonicks: Sol para nuestros Ancianos. = The second project, Valorización de la Quínoa Atacameña a través de la caracterización nutracéutica, elaboración de productos funcionales y cadenas de comercialización, was funded through the Agricultural Innovation Fund and sponsored by the Comisión Nacional de Investigación Científica y Tecnológica. This project will bolster cultural ESs related to traditional crops and techniques.

Policy impact

One of ProEcoServ-CL's strategies to mainstream ESs into policy support was to identify and link existing high-level policy support tools (such as international conventions, laws of the republic, regulations, political strategies and national plans) with other tools operating at the regional and local level. At the regional level, ProEcoServ-CL identified two main tools that included ESs in the policy and decision making: Regional Development Strategy and the Action Plan for Biodiversity in the Antofagasta Region. At the local level, ProEcoServ-CL identified the Communal Development Plan and Municipal Ordinance for Local Environmental

Management. It is worth noting that ESs are not clearly integrated into this political and legal framework, although the ecosystem concept is mentioned in the higher-level policy tools.

In this context, the training and dissemination of Tableau software was critical to enabling policy makers and stakeholders to use the ProEcoServ-CL results to support decisions. During the change in government in March 2014, the Municipality of San Pedro de Atacama experienced shifts in the local policies aimed at protecting and managing ecosystems in a sustainable way. Furthermore, presenting the satellite-based water balance for the Central Altiplano proved that highly sophisticated scientific tools and approaches can be distilled into one simple figure that can still drive home a powerful point. In the SPA case, it allowed the discussion to move from speculation to fact-finding. This has never been achieved in the current national context of broad mistrust of established authorities, which suggests a way forward for dialogue around ESs.

Another concrete policy-making impact is that the Municipality of San Pedro de Atacama is actively working on the first-ever tourism development plan for the Comuna (PLADETUR). They must first determine which area will be assigned a special zoning status designed for tourism development (ZOIT). To achieve this objective, decision makers are using the Tableau software platform to visualize the tourism data that has been collected and systematized by the ProEcoServ-CL team during the project. Furthermore, representatives from the local office of the National Forestry Corporation (CONAF) and of the National Service of Tourism (SERNATUR) have learned from Tableau how to interpret tourism dynamics and how they can affect ecosystems and tourism attractions, and how the water provision between seasons determines the water availability for local communities and tourism activity. Both activities are interesting achievements, as they highlight the potential for engaging with local-level actors in the design of policies that have an enormous potential to cascade into other local initiatives across the country. Given the very limited room for compromise that currently exists between local development and national authorities, particularly where indigenous groups are involved in decision making, these achievements are providing a nationwide benchmark for how to design effective ES policies.

Another important aspect of ProEcoServ-CL refers to the inclusion of ESs in regional policy and national legal tools. This objective was, however, solidly based on local assessment of environmental policies, scenario development for SPA, and application of different strategies for economic valuation of ESs (surveys, workshops and meetings). Each of these actions was carried out during the ProEcoServ-CL project execution in order to influence national policy making. Naturally, this work approach also considered the local relevance of such actions, plans and policies. Without this consideration, the impact that an initiative could have on national policy could not contribute in the same way to the welfare of local communities and their ecosystems. Therefore, a constant exchange of knowledge is necessary between the different levels of decision makers to strengthen the ecosystem approach in the development of the country.

Probably one of the clearest indicators of this ESs integration assessment and political engagement are the projects in which CEAZA has been registered during 2015, ProEcoServ-CL's last working year. Two projects awarded have included the local criteria in the proposals using regional (Regional Ministry of Environment, or FPA) and national (Ministry of Agriculture, or FIA) funding. The implication is that ProEcoServ-CL results have been internalized to support the continuous integration of ESs criteria into the national policy, considering the local relevance as the framework. At this stage, it is also worth mentioning two key results that emerged from the installation of the ProEcoServ-CL team in SPA. Firstly, CEAZA has access to a very good network of professionals that are interested in the sort of applied science activities that CEAZA scientists conduct. Therefore, the ecosystem concept, and the whole concept of evidence-based decision making around ESs, as well as the direct benefits a long-term view may bring about to local communities, is now firmly established in SPA. Secondly, the Council of Atacameñan People (Consejo de Pueblos Atacameños, CPA) was explicitly included in a high-level presidential consulting body that delivered a major policy white paper on the future of lithium mining in Chile, which is now centered in SPA. This policy document explicitly and profusely includes the ecosystem concept, water provisioning, eco-tourism and the need for sustainable development to preserve fragile ESs that are key to local livelihoods, representing very concrete proof of the national-level impact of ProEcoServ-CL's work.

To summarize, the policy impact can be assessed in different ways and perspectives. There is a lack of state-level tools to project decision making from the national to local level. Therefore, the ProEcoServ-CL team chose, as a core strategy, to focus on the local work to accomplish this objective. The work was focused on sectoral strategies to achieve a bottom-up influence on higher-level ESs in policy making. Our results show that complex initiatives, such as ProEcoServ-CL, can only be developed at the local (municipal) scale if they are validated with decision makers within the local community (e.g. CPA and Municipality of SPA), and supported by national-level institutions (Ministry of the Environment, Ministry of Agriculture). In this sense, ProEcoServ-CL provides a clear and important precedent for ongoing efforts to operationalize ESs, both in SPA and Chile generally.

Conclusions and recommendations

ProEcoServ-CL's work suggests a series of strategic and practical recommendations to ensure a better understanding of, and capacity building for, sustainable management of ESs in San Pedro de Atacama. To ensure rigorous scientific assessment, modelling and valuation of ESs, it is still necessary to establish real-time monitoring networks to better understand ES dynamics and improve future scenarios for the region and the municipality. Thus, implementing an observation platform with continuous measurements, providing near-real time information of ESs throughout Internet and social media platforms, will enhance decision support capabilities and management of ESs in areas under study. In this respect,

ProEcoServ-CL has made a significant contribution to developing the social capital that must precede the construction of physical infrastructure (hydrological and meteorological stations, groundwater measurements or monitoring visitors in real-time) to nurture sustainability as communities prepare to assimilate these key pieces of information. Furthermore, ProEcoServ-CL established local leaders and participatory processes that can provide interesting technical options. Managing and designing physical ES infrastructure in a participatory way can be replicable in other projects with the tools that every country has available. Using this methodology, specific solutions to ensure human welfare and better management of ESs for each country can emerge. Implementation of the Tableau software platform in the local administration of SPA (Municipality of SPA, Council Atacameños People, CONAF, SERNATUR) is a concrete example in this regard. To our knowledge, ProEcoServ-CL is the first example in this country of a participatory process that has successfully empowered and matched indigenous people with the ESs concept managed by national authorities. Discontent among indigenous people in Chile is increasingly centered on access to ESs, and there is a lack of tools to match the contrasting perception held by authorities and indigenous leaders in this regard. Therefore, the use of this participatory process is under consideration in other contexts, such as Easter Island, where CEAZA scientists have a strong local presence. Similarly, ProEcoServ-CL has submitted a proposal to the regional government of Antofagasta, where the SPA indigenous community will participate in the monitoring effort highlighted at the beginning of this chapter, with the intention of creating a permanent regional study center focused on ESs sustainability in the broader Altiplano region.

Based on the ProEcoServ-CL San Pedro de Atacama experience, our recommendation is to strengthen and monitor the bridge between science and policy throughout participatory processes at different levels. This must be done throughout the entire project, emphasizing the evaluation stage and the design of products (deliverables). As noted previously, implementing agencies (i.e. CEAZA), should be provided with higher-level counterparts within the central administrations that are in more powerful technical positions and not politically appointed. It was a huge burden to reinitiate dialogue with low-level officials who held little political influence in the central administration, and who did not clearly understand the aim of the proposal. Although resources and personnel were dedicated to train them, it was of little practical use, since their requests or motivations were dictated from a political vantage point, with no room for the compromise that ESs mainstreaming required.

References

Casado-Arzuaga I., Onainda M., Madariaga I., Verburg P.H. (2013) Mapping recreation and aesthetic value of ecosystems in the Bilbao Metropolitan Greenbelt (northern Spain) to support landscape planning. *Landscape Ecology*. DOI: 10.1007/s10980-013-9945-1
Cowling R.M., Egoh B., Knight A.T., O'Farrell P.J., Reyers B., Rouget M., Roux D.J., Welz A., Wilhelm-Rechman A. (2008) An operational model for mainstreaming ecosystem

services for implementation. Proceedings of the National *Academy* of *Sciences USA* 105:9483–9488.

Daily G.C., Polasky S., Goldstein J., Kareiva P.M., Mooney H.A., Pejchar L., Ricketts T.H., Salzman J., Shallenberger R. (2009) Ecosystem services in decision making: Time to deliver. *Frontiers in Ecology and the Environment* 7(1):21–28.

Dirección General de Aguas (2015) *Información Oficial Hidrometeorológica y de Calidad de Aguas en Línea.* (www.dga.cl/servicioshidrometeorologicos/Paginas/default.aspx)

EuroChile (2006) *Proyecto Innova Chile Corfo – Puesta en Marcha de un sistema de gestión ecoturística sustentable para el destino territorial San Pedro de Atacama 2005–2006.* Santiago: EuroChile.

Global Synthesis Report of the Project for Ecosystem Services (2015) UNEP, Ecosystem Services Economics Unit, Division of Environmental Policy Implementation.

Hernandez-Morcillo M., Plieninger T., Bieling C. (2013) An empirical review of cultural ecosystem service indicators. *Ecological Indicators* 29:434–444.

Hiebert B. (2012) *Post-Pre Assessment: An Innovative Way for Documenting Client Change.* Department of Educational Psychology and Leadership Studies. University of Victoria, BC.

Instituto Nacional de Estadisticas (INE) (2014) EMAT, Informe Annual. National Tourism Service. Santiago.

Longuevergne L., Wilson C.R., Scanlon B.R., Crétaux J.F. (2013) GRACE water storage estimates for the Middle East and other regions with significant reservoir and lake storage. *Hydrology and Earth System Sciences* 17:4817–4830.

McCool S.F., Clark R.N., Stankey G.H. (2007) *An Assessment of Frameworks Useful for Public Land Recreation Planning.* General Technical Report PNW-GTR-705. Portland, OR: US Department of Agriculture, Forest Service, Pacific Northwest Research Station. 125 p. 705.

Millennium Ecosystem Assessment (MA) (2005) *Ecosystems and Human Well-Being: Synthesis.* Washington, DC: Island Press.

Nahuelhual L., Carmona A., Lozada P., Jaramillo A., Aguayo M. (2013) Mapping recreation and ecotourism as a cultural ecosystem service: An application at the local level in Southern Chile. *Applied Geography* 40:71–82.

Peña L., Casado-Arzuaga I., Onainda M. (2015) Mapping recreation supply and demand using an ecological and a social evaluation approach. *Ecosystem Services* 13:108–118.

Saaty T.L. (2008) Decision making with the analytic hierarchy process. *International Journal of Services Sciences.* DOI: 10.1504/IJSSci.2008.01759

Sitas N., Prozesky H.E., Esler K.J., Reyers B. (2013) Opportunities and challenges for mainstreaming ecosystem services in development planning: Perspectives from a landscape level. *Landscape Ecology.* DOI: 10.1007/s10980-013-9952-3

Tapley B., Ries J., Bettadpur S., Chambers D., Cheng M., Condi F., Gunter B., Kang Z., Nagel P., Pastor R., Pekker T., Poole S., Wang F. (2005) GGM02: An improved Earth gravity field model from GRACE. *Journal of Geodesy* 79:467–478. DOI: 10.1007/s00190-005-0480-z

Wood S.A., Guerry A.D., Silver J.M., Lacayo M. (2013) Using social media to quantify nature-based tourism and recreation. *Scientific Reports* 3:2976. DOI: 10.1038/srep02976

7 Integrating natural capital and ecosystem services into policy and decision making in Trinidad and Tobago

John Agard, Lena Dempewolf, Maurice Andres Rawlins, Carl Obst, Carlos Muñoz Brenes, Shirley Murillo Ulate and Keisha Garcia

Introduction

As a small island developing country, Trinidad and Tobago's national economic activity, environmental security, and human health crucially depend on the country's rich biome and diverse ecosystems. And thus, the mandate of ProEco-Serv-Trinidad and Tobago (ProEcoServ-T&T) was to focus on mainstreaming ecosystem services into decision making via policy, legislation, national accounting, development planning and a payment for ecosystem services (PES) scheme. This chapter will discuss the following interventions: (1) the development of ecosystem value maps for spatial development planning; (2) the integration of natural capital accounting, such as of carbon, water, land use and cover and biodiversity into the national accounts; and (3) the piloting of a payment for ecosystem services model to foster the sustainable use of valuable ecosystems. The work was coordinated by the University of the West Indies (UWI).

Ecosystem value maps for spatial development planning

The development and introduction of ecosystem value maps for spatial development planning was piloted for selected ecosystem services that were retained, after stakeholder consultation, as Trinidad and Tobago's key ecosystems in terms of their role and contribution toward the support of local livelihoods. These refer to the forests in the Northern Range and their contribution to sediment retention; the Nariva swamp, freshwater wetland and designated a Wetland of International Importance under the Ramsar Convention, and its contribution in terms of carbon sequestration and pollination services to small farmers.

The Northern Range: valuing soil retention services provided by forests

The Northern Range presents the most dominant relief feature and ecosystem on the island of Trinidad. Covering approximately 25% of Trinidad's land mass,

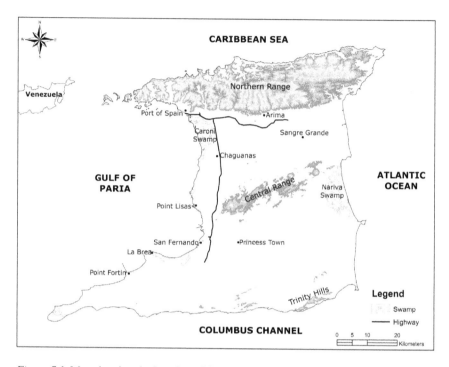

Figure 7.1 Map showing the location of the Northern Range area of Trinidad and Tobago.

this range is covered by a variety of tropical forest, primarily seasonal evergreen tropical forest, also with semi-evergreen seasonal forest, deciduous seasonal forest, dry evergreen forest and seasonal montane forest. This range runs in a west-to-east direction along Trinidad and Tobago's north coast and is an ever-present feature in the lives of the residents along Trinidad's east-west corridor, the main population corridor of Trinidad located along the foothills of the entire range (Figure 7.1).

The Northern Range provides a wide variety of provisioning forest ecosystem services to residents of Trinidad and Tobago, including recreational hunting, timber and non-timber forest products for medicine and craft and water. As a mountainous small island state, Trinidad inevitably faces flooding issues, which not only result in rapid runoff in extreme rainfall periods but encourage development of flatter flood-prone areas. Poor management of the Northern Range forests results in the degradation of a key ecosystem service, soil erosion and soil retention. The loss of this ecosystem service results in the degradation of soil quality in agricultural areas as well as in downstream or offsite effects, including the increase of the risk of river flooding and siltation of dams.[1]

For the valuation of forests' soil retention services, the proposed method was characterized by a biophysical assessment where rates of potential soil erosion and soil loss was modelled using the Revised Universal Soil Loss Equation (RUSLE) (Yoder et. al., 2001):

$$A = R * K * LS * C * P$$

Where:

A calculated soil loss through erosion from agricultural plots in the Northern Range
R rainfall factor determined by rainfall intensity over a 12 month period
K factor based on soil properties such as soil fraction or the relative clay, sand and silt content of soil
LS factor based on the length and angle of slope
C factor based on land cover
P factor based on local conservation practices

This model was coupled with data from field experiments to validate this locally applicable model. Soil loss data was obtained from erosion plots established in the Maracas watershed by the Forestry Division. Erosion plots were characterised by different land covers including forest, annual crops (agriculture), grassland, and degraded forest. Erosion plots were 10m x 5m bounded plots, and erosion data was scaled up from values in units of g m^{-2} yr^{-1} to units of t ha + yr^{-1}. Slopes of the estimated erosion values are consistent with the slope of the observation erosion plots. Model validation was undertaken for the years 2007–2008 using descriptive statistics, and it was found that the model performed satisfactorily.

Following validation of the model, it was used in ArcGIS to model erosion on a catchment basis using input data for the years 1993, 1995 and 1998–2002 (Rawlins, 2017). The ArcGIS platform allowed for the model to be applied adjusting for the changes in the **K**, **LS**, **C** and **P** factors on the finest spatial scale of data available for improved accuracy. This method produced results on the volume of sediment production or sediment retention for each hectare of the Northern Range. Table 7.1 presents the estimation results.

As can be seen from Table 7.1, forests retain a significant volume of soil across all watersheds in the Northern Range. On an annual basis, the Northern Range prevents erosion of about 6.7 million tonnes of sediment, which is valuable for agriculture.[2] As the use of purchased topsoil is a practice familiar to farmers to mitigate soil erosion impacts, it is proposed to assess the value of this erosion prevention based on the market value for topsoil. This economic valuation method is known in the literature as the replacement cost method and it calculates the financial costs that would be incurred to replace the service provided by a damaged or removed natural asset, in this case topsoil. Topsoil in Trinidad and Tobago was being sold at prices ranging from US$82.20 per tonne when sold by small suppliers in 100-pound bags to US$19.20 per tonne when sold per truckload by large suppliers with transportation costs included.

Table 7.1 Summary of results from biophysical assessment of erosion prevention services provided by Northern Range forests

Watershed Name	Watershed area (ha)	Sediment exported from watershed (tonnes)	Predicted erosion (tonnes)	Eroded sediment retained due to forest (tonnes)	Sediment export ratio
Tocco	20,920	14,821	598,764	479,644	2.48%
Yarra	4,144	512	104,593	104,265	0.49%
Madamas	8,298	17	145,963	146,810	0.01%
Marianne	4,734	22	13,571	12,690	0.16%
Rest North	4,614	–	115,613	105,464	–
North Oropouche	7,354	56,817	1,586,669	1,534,457	3.58%
Salybia	6,040	13	443,094	369,732	0.00%
Matura	5,267	16,303	788,478	828,189	2.07%
Chaguaramas	10,144	21,107	700,987	661,181	3.01%
Santa Cruz	6,645	11,336	238,694	208,004	4.75%
Maraval	2,570	1	234,142	236,929	0.00%
Maracas	4,886	6,373	31,026	69,451	20.54%
Caura/Tacarigua	4,885	22,986	270,491	247,614	8.50%
Guanapo	4,949	9,887	121,909	106,253	8.11%
Arima	4,414	5,993	173,877	177,960	3.45%
Hollis	1,420	–	807	749	–
Arouca	5,947	3,631	689,039	684,911	0.53%
Aripo	5,248	9,877	281,652	275,641	3.51%
Quare	9,208	88,972	336,986	241,522	26.40%
Rest North Oropouche 1	6,609	871	110,414	99,412	0.79%
Port-of-Spain	3,881	98	80,169	86,870	0.12%
Tunapuna	1,773	43,553	97,822	8,665	44.52%
Mausica	2,125	–	31,708	30,053	–
El Mamo	2,242	1,907	44,714	36,534	4.27%
Orupuna	2,003	–	20,357	19,497	0.00%
Total		**315,097**	**7,261,538**	**6,772,497**	**4.34%**

Source: Author's computation

In this context, the economic value of soil eroded in a year is calculated as:

$$VSE = A * PTS$$

Where:

A   calculated soil loss through erosion from agricultural plots in the northern range

PTS average national price for topsoil most similar in composition to naturally occurring soil in the NR

Taking into account these market prices, the economic value of soil retention services provided by the Northern Range forests ranges between US$374 and 622 million (Table 7.2). This value is quite significant for the national economy of Trinidad and Tobago as it represents about 4.0% to 6.8% of the central government annual revenues, reiterating how important is to have this ecosystem service factored in a national planning policy for Trinidad and Tobago. In this context, ProEcoServ-T&T, which was launched by the ministry responsible for planning,

Table 7.2 Summary of economic value of erosion prevention services provided by Northern Range forests (on an annual basis)

Watershed Name	Eroded Sediment retained due to forest tonnes)	Replacement cost		
		Lower value (million US$)	Upper value (million US$)	Value of erosion prevention (US$/ha/year)
Tocco	479,644	26.5	44	1,306
Yarra	104,265	6.5	9.5	1,391
Madamas	146,810	8.17	13.5	978
Marianne	12,690	0.67	1.17	148
Rest North	105,464	5.83	9.67	1,263
North Oropouche	1,534,457	84.83	141	11,531
Salybia	369,732	20.5	34	3,383
Matura	828,189	45.83	76	8,689
Chaguaramas	661,181	36.5	60.67	3,602
Santa Cruz	208,004	11.5	19.17	1,730
Maraval	236,929	13.17	21.83	5,094
Maracas	69,451	3.83	6.33	786
Caura/Tacarigua	247,614	13.67	22.67	2,801
Guanapo	106,253	5.83	9.83	1,187
Arima	177,960	9.83	16.33	2,228
Hollis	749	0	0	29
Arouca	684,911	37.83	62.83	6,365
Aripo	275,641	15.17	25.33	2,903
Quare	241,522	13.33	22.17	1,450
Rest North Oropouche 1	99,412	5.5	9.17	831
Port-of-Spain	86,870	4.83	8	1,237
Tunapuna	8,665	0.5	0.83	270
Mausica	30,053	1.67	2.83	782
El Mamo	36,534	2	3.33	901
Orupuna	19,497	1	1.83	538
Total	**6,772,497**	**374.99**	**621.99**	**2,457**

Source: Author's computation

was appointed by the government to be a member of the Development Planning Steering Committee of Trinidad and Tobago. In this context, ProEcoServ had an opportunity to have a policy impact on the Planning and Facilitation of Development of Land Bill before the Parliament.

The Nariva Swamp area: valuing pollination services to small farmers

Another key ecosystem service that ProEcoServ-T&T submitted for economic valuation is the pollination services that wild bees and insects provide to small farmers. The valuation study was implemented in the Nariva Swamp, as this swamp and surrounding areas is sounded by a vast agricultural area upon which depend a variety of crops for subsistence and vegetable cash crops such as cucumbers. The pollinators support these agricultural activities, impacting the total levels of production and thus impacting the livelihoods of the locals. This ecosystem service is of particular importance as this is one of the least developed areas of Trinidad and Tobago with the lowest levels of income per capita – about US$5,000 per year. With the residents' comparatively low incomes, the economic and subsistence activities supported by Nariva Swamp, such as agriculture, hunting, fishing and crab and shellfish gathering, are of comparatively higher importance.

For the valuation of wild pollination services to agricultural goods produced by the local rural communities in the Nariva Swamp, a production function was used, where the contribution of pollination to crop **A** is modelled as follows:

$$Q = K * L * Land * D$$

Where:

Q   production of crop A
K   amount of capital inputs used in the production of crop A
L   amount of labour inputs used in the production of crop A
L   **and** amount of land inputs used in the production of crop A
D   insect dependent factor in the production of crop A

The method to estimate insect dependent factors in the production of a crop entails a series of plots which are placed at varying distances from the edge of the swamp, covered by nets of varying sizes. This works to exclude pollinators of certain sizes from crops to remove their influence on crop pollination and, hence, yield so that pollinator influence could be disaggregated. To assess the marginal impact of wild pollination on crops, this experiment was replicated at four different controlled experiment sites: Plot I at each location was covered with ¾" wire mesh to exclude insects with a body width larger than ¾" in diameter; Plot II at each location was covered with ¼" wire mesh in order to achieve the same with respect to insect exclusion of this size class; flower buds of plants in Plot III were covered with mesh bags to exclude all insects; the open Plot IV remained uncovered. All insects were free to access flowers in this plot. Each plot measures approximately 4.9 m × 2.4 m, equalling an area of 11.9 m². Crop yields and seed yields are then recorded and examined to determine the relative influence and importance of pollination services provided by insects that rely on the Nariva Swamp as a habitat to the agricultural activities in the areas surrounding the swamp. Cucumbers were selected as the primary crop of study as they were identified as key crops for exploration and listed as a priority commodity in the National Food Production Action Plan 2011–2014. The insect pollination results across showed that the insect dependence factor D was approximately 0.965 for cucumbers and 0.769 for hot peppers.

This resulted in a very significant mean percentage reduction in yield in the complete absence of pollinators, ranging from 76.9% to 96.5%, for hot peppers and cucumbers, respectively. The loss of income as a consequence of total exclusion of pollinators from flowers to farmers growing hot peppers would potentially incur weekly losses ranging from US$398 to 861 per acre, while losses per acre of cucumbers amounts to between US$2,348 and 12,692 per crop cycle, which is approximately eight weeks (Dempewolf, 2018).

Finally, we explore scaling up economic valuation of the pollination results from the ProEcoServ-T&T study to the country's national economy. The value of agriculture output attributed to insect pollination is calculated as follows:

$$V_{ip} = (D_x * V_x)$$

Where:

V_{ip} annual value of agricultural output attributable to insect pollination
V_x the annual value of crop X as produced in Nariva
D_x the insect dependence factor of crop X as calculated by ProEcoServ

The insect dependence factors developed in Nariva were multiplied by the total annual value of produce nationally of selected vegetables, so as to calculate the proportion of the annual value of crops produced attributable to pollination. The total value of these crops was determined by multiplying average annual prices and national production data, both provided by the Ministry of Food Production. Table 7.3 summarizes the results. Lower and upper value estimates indicate two pollination scenarios: in the complete absence of pollinators, and a scenario where the activity of pollinators was reduced to 25% of the current scenario. For example, in 2012 the annual production of cucumbers was 1,355 tonnes, generating a revenue of US$1,059,209. Taking into account the results from the field experiments, the marginal value of the insect pollination to the selected crops ranges between US$193,387 and 1,022,121. In other words, the scenario of total absence of pollination is associated with a significant expected economic loss, in which the upper value is almost the same as the recorded production. From the national economy viewpoint, the economic value of pollination can be estimated to be in the range of 9% to 13% of the annual value of vegetable production in Trinidad and Tobago.

Table 7.3 Economic value of pollination of cucumbers and hot peppers (in US$)

	2010	2011	2012
Cucumber Production (tonnes)	1,300	1,193	1,355
Market value (USD)	838,279	1,054,744	1,059,209
Marginal value of pollination (USD)	153,699–808,940	193,387–1,017,829	193,387–1,022,121
Hot pepper Production (tonnes)	700	491	250
Market value (USD)	3,466,400	2,468,297	2,837,051
Marginal value of pollination (USD)	1,226,204–2,665,662	873,135–1,898,120	1,003,578–2,181,692

Source: Author's computation

Given that agriculture is important for income generation as well as subsistence crops for the communities of Nariva, pollination is key for supporting a governmental policy targeted at improving food security. The results also show that the impacts of a change in the pollination conditions differ across the two selected vegetables. Though the wild pollination services studied here are valued primarily as an input in agricultural production, it is important to note that pollination services are also critical to ecosystem functioning. Specifically, they support the propagation and maintenance of key fauna and flora that provide critical habitat and regulatory services that support the delivery of all other services provided by the swamp. The valuation of these ecosystem services is beyond the scope of the present study.

The Nariva Swamp is also key to the reduction of coastal erosion in the area it is located. The roots and stems of mangroves work to dissipate the energy of the impact of coastal waves, which reduces their ability to erode and transport sediments and other materials. This induces the process of deposition, contributing to a feedback loop whereby mangroves can further settle and stabilize existing sediment deposits. Due to its direction towards the open ocean, Trinidad's west coast encounters significant coastal erosion issues. Thus, the location of a number of key mangrove forests, namely the Nariva Swamp, the Fishing Pond Mangroves, Matura Bay Mangroves, Point Galeota Mangroves and Guayaguayare Mangroves, is vitally important to reducing erosion rates along Trinidad's west coast. The importance of these coastal protection services must be emphasised in the context of sea level rise associated with climate change. Mean sea level is expected to rise by 0.35 m globally by the end of the century in the A1B scenario (IPCC, 2007).[3] This rise is expected to vary greatly from region to region and thus the Caribbean, and Trinidad and Tobago, may face even greater rates of rise.

Nariva is home to 45 mammal species, 30 reptile species, 33 fish species, 204 bird species, 19 frog species, 213 insect species and 15 mollusc species. From an economic standpoint, some of this biodiversity value of this ecosystem is captured in the value of commercially exploited wildlife or other activities such as tourism that are supported by the presence of species present in the swamp. Some juvenile fish species spend a portion of their life cycle in mangrove forests, within which they have a much higher survivorship rate compared to the open sea. Hence, mangrove forests have been shown to enhance the biomass of offshore Caribbean reefs because of the functional linking between the two ecosystems (Mumby et al., 2004).[4]

Finally, the ProEcoServ team investigated the potential of carbon sequestration of the Nariva Swamp. According to computations of the ProEcoServ country team, the average amount of carbon sequestration in this area is estimated at 135 tonnes per hectare per year. This corresponds to annual carbon sequestration service of 1.53 million tonnes of carbon a year, which equates to 11.1% of all CO_2 emissions from Trinidad and Tobago.[5] In other words, the Nariva Swamp plays a key role as a nature-based solution to stock carbon, and this ecosystem-based tool for carbon sequestration has a great financial potential, including PES for a carbon-based management scheme.

Conclusions

This section provides an overview of the significant economic value of selected ecosystems services (UNEP, 2014), including pollination and soil retention services provided by forests in Trinidad and Tobago. Furthermore, it highlighted the importance of choosing a valuation method most appropriate for the benefit that is to be valued. Finally, economic valuation results inform the policy maker of the importance in incorporating ecosystem services in spatial development planning. The failure to incorporate this dimension may result in increased risk of natural disasters such as flash floods, as well as a loss of agricultural productivity through factors such as soil erosion

Developing a natural capital accounting framework for Trinidad and Tobago

Setting the scene

The key objective for this aspect of the project was to integrate natural capital accounting with the standard national accounts so as to broaden the information available to decision makers when considering macro-economic and national sustainable development and planning issues. The accounting framework is therefore a tool rather than an end in itself. This point is important to recognize since the full integration of natural capital accounts is a challenging task both conceptually and practically. The project identified a range of options towards full integration that can be described, with each option broadening the available information set. These options emerge from considering the implementation of the adopted UN standard – the System of Environmental Economic Accounting (SEEA) that is now the agreed framework for national level accounting (United Nations et al., 2014a, 2014b). The four options identified were:

Option 1: Assessing expenditure on environmental protection or resource management with outcomes in terms of changing ecosystem conditions.

The ongoing measurement of ecosystem conditions across the country enables a more accurate evaluation of the value of expenditure related to policy initiatives in bringing about positive environmental outcomes. The objective is to change the determination of policy success away from being determined solely on traditional measures such as number of people employed and number of dollars spent, etc.

Option 2: Combined accounts for certain economic activities.

A combined account is a structured table that brings together data in monetary terms (e.g. output, value added, investment) with data in physical terms (e.g. water use, energy, use, emissions, etc.) to provide a more comprehensive picture of a particular economic activity. Combined accounts are shown in Chapter 6 of the System of Environmental Economic Accounting Central Framework (SEEA-CF), relating to energy, forest products, water and emissions. Also possible are combined accounts for agricultural activity (SEEA-AGRI which has been developed by FAO) and for tourism activity.

Option 3: Balance sheets for natural resources and depletion-adjusted GDP.

The SEEA Central Framework describes the way in which standard balance sheets including produced (built) capital and financial capital can be extended to incorporate the value of natural resources. Using these balance sheet values, estimates can then be made for the depletion of natural resources, for example depletion of mineral and energy resources, and measures of GDP can be adjusted for depletion. (Note that this adjustment does not provide a complete adjustment for the cost of using up environmental assets, and only includes the costs associated with the extraction of natural resources.)

Option 4: Recognizing the full value of ecosystem assets and their services.

The SEEA Experimental Ecosystem Accounting describes an integration approach whereby ecosystems can be added onto the standard accounting structure to reflect an additional "sector" with which other sectors, such as corporations, government and households, interact. The accounting would show flows of ecosystem services between the ecosystem sector and economic sectors, as well as show declines in the value of the ecosystem sector due to degradation of the ecosystem assets. The addition of another "column" in the accounting structure is difficult in practice but is a powerful organizing concept (UNEP, 2014).[6]

After considering these options, and in the light of the policy drivers in Trinidad and Tobago), the project proceeded to: (1) describe the main actions required to implement natural capital accounts ("Mangrove provisioning services"); (2) design and test several demonstration accounts for particular aspects of natural capital ("Coastal protection, carbon sequestration benefits (mangrove regulating services)"; and (3) assess the potential for implementation in terms of factors supporting the development of natural capital accounting in T&T ("Factors supporting implementation in Trinidad and Tobago"). The findings on these issues are presented in the following sections with some concluding remarks in "Conclusions".

Plan of action for implementing natural capital accounting

The starting point for the development of national natural capital accounts for Trinidad and Tobago is to provide a comprehensive assessment of the country and its environment. This may appear to be a challenging task but can be separated into manageable stages in which more detail may be progressively incorporated into information sets.

In Trinidad and Tobago, key steps can be taken to allow for the implementation of the approach described previously. The proposed steps are congruent with advice presented by the international statistical community for the implementation of the SEEA (see United Nations, 2013; SEEA Implementation Guide) and these steps are also being tested for ecosystem accounting in several pilot countries.

Step 1: Establish a core group

The core group should be a high-level cross-agency group involving both potential users and producers of natural capital accounting information. This group should involve representatives from central agencies and is intended to provide direction and support to technical work.

Step 2: Strategic planning

Under the guidance of the core group, a small team should undertake a planning process aimed at establishing the necessary relationships, legislative requirements, data sharing commitments, staffing and resources, and accounting priorities. The focus here should be on medium- to long-term planning while identifying areas in which short-term results may be found. Since the development of accounting experience is not a one-off exercise but requires a sustained commitment of at least three to five years, establishing a sense of continuity is important.

Step 3: Involvement of international agencies and regional bodies

To ensure success, the project must be led by Trinidad and Tobago. At the same time, there is ever-increasing support for natural capital accounting within the international community – both statistical and otherwise – and connections to relevant networks are likely to be extremely advantageous in supporting the establishment of a natural capital accounting program at the country level. Further, as more countries within the region develop natural capital accounting programs, opportunities to exchange experience, knowledge and techniques should be taken.

Step 4: Priority accounts

Notwithstanding the outputs from the strategic planning process in Step 2, it is proposed that initial focus be placed on developing land accounts and water accounts. This proposal is based on the requirements for policy, the availability of data, and the emerging logic of structuring natural capital accounting programs. In essence, the information from the two accounts will likely provide the most effective broad-level assessment of changes in environmental condition. Of note is the importance of tracking area of forests when conducting land accounts.

Step 5: Establish core natural capital accounting team

One key lesson from countries that have established programs on natural capital accounting is that it is essential to start small and then build. Part of the challenge is gathering data, but a bigger challenge lies in establishing the discipline of accounting, which requires not only obtaining the appropriate data but organizing, confronting, reconciling and balancing these data to provide a coherent story. Accounting is, in essence, a "learning by doing" exercise, so it is helpful to begin with relatively basic accounting structures that develop in complexity over time. Presenting initial efforts as experimental or demonstration accounts is a standard practice in a number of countries and is an excellent way to generate interest, garner feedback and establish networks.

Step 6: Ensure allocation of resources for the communication and use of accounts

While the release of accounts is a significant achievement, just as important is that on completion of accounts there is widespread communication, explanation and promotion of the accounts and process. It is also recommended that explicit efforts are made to work with governments to understand how the accounts can be used in decision-making situations. The collaboration between producers and users is likely to be invaluable in ensuring that the accounts are appropriately adapted and relevant.

With sufficient commitment and resourcing (four to six people ongoing), the establishment of a natural capital accounting program should generate results within 12–18 months. A number of different agencies are well positioned to lead the work and effectively "house" a natural capital accounting program. A close connection with the National Accounts Department of the Central Statistical Office is essential – or the office should even host the program itself. Alternatively, a new unit in the Ministry of Planning and Sustainable Development may be an effective option, or an academic institute could compile the accounts, to be subsequently reviewed by the Central Statistics Office (CSO), as is the case in Guatemala. It is noted that it is not sufficient for the data to be released by an academic unit – government endorsement and support is crucial.

Demonstration accounts for carbon, land, water and biodiversity

As part of this project, four "demonstration" accounts were designed and implementation tested. The intention was to provide a general indication of what some natural capital accounts might look like and to assess what type of information may be readily available. The four accounts were for: carbon stocks (covering mineral resources and forests), land cover, water resources and species diversity. In the sections that follow, the chapter discusses key factors surrounding the accounting for the select natural resources and further discusses aspects of relevance for policy and monitoring for Trinidad and Tobago.

Carbon

Carbon accounting usually refers to the measurement of emissions of greenhouse gases that arise as a result of economic activity, though the broader ambition is to track the stocks and flows of carbon between the different places that carbon is stored.

Ideally, a carbon stock account would cover all stores/reservoirs of carbon within a country, whether as geo-carbon or bio-carbon. Geo-carbon includes reserves of oil, gas, coal and similar resources and also carbon held in rocks, primarily limestone and some minerals (e.g. marine sediments and methane clathrates). Bio-carbon covers carbon stored in all plant and animal species (living and dead) and in soil and water resources. From a national perspective, carbon stored in the oceans and the atmosphere is generally not allocated to a country and, as appropriate, would instead be incorporated into global models of accounting.

The relevance of carbon stock accounting to Trinidad and Tobago is important for two main reasons. Firstly, carbon is an important key indicator for assessing changes in ecosystem conditions and thereby working towards integrating natural capital into economic accounts. Secondly, given the relative importance of Trinidad and Tobago's oil and gas industry, understanding the available stores of carbon in the oil and gas reserves and changes in those stores represents an important aspect of integrating measures of natural capital with economic activity. The oil and gas sector in Trinidad and Tobago is of clear significance, accounting for 43% of the country's GDP directly, and supporting a range of other activities (particularly manufacturing) through the availability of a readily available supply of energy. Since oil and gas are non-renewable resources, assessing their long-term impact on national income requires an understanding of the status of total reserves and the feasibility of their abstraction over time. The annual report commissioned by the Trinidad and Tobago government (Ryder Scott Report) on the status of the country's oil and gas reserves is recognition of the need for these data. Also, the developing work of the Trinidad and Tobago Extractive Industries Transparency Initiative (TTEITI) to ensure consistent reporting of oil- and gas-related flows between government and corporate entities is an important activity in public accountability for the resources. These two reporting activities could be supported by a general programme of natural capital accounting for sub-soil resources following the international statistical standard – SEEA 2012 Central Framework (United Nations et al., 2014a). Reporting following the SEEA would ensure coherence in accounting for the oil and gas resources with the measures of GDP and other measures of economic activity as well as allowing for comparison to other countries. At a macro level, SEEA-based reporting provides a framework for assessing long-term extraction trends and asset-life in a way that enables comparison between different natural resources and different countries. One of the demonstration accounts, the carbon account, also shows how estimates of oil and gas reserves may be converted into carbon equivalent measures and hence combined with other sources of carbon, primarily forests, to create a broader carbon stock account.

Techniques for the measurement of the carbon in these stores will be discussed with experts in Trinidad and Tobago, utilizing experience in the development of carbon stock accounts in Australia, and taking into account the measurement of changes in carbon stock used in the ecosystem condition measures developed by the European Environment Agency.

Land cover

A basic requirement for natural capital accounting is an understanding of how a country's land cover has changed over time. Land cover accounts use a broad classification of types of land cover – e.g. forests, wetlands, agricultural land, inland waters, urban areas, etc. – to record the number of hectares of each land cover type at various points in time. From here, it is possible to analyse the changes between points in time examining additions and reductions in each land cover type and classifying the changes by whether they are driven by economic or perhaps other causes. The possibility of using GIS and satellite data should be investigated in the first instance when developing land cover accounts. Trinidad and Tobago may already have land cover maps and classification developed by different agencies for specific purposes (e.g. management of water resources) so it would be advantageous to seek first what is available followed by concordance to proposed international classifications in the SEEA.

Ideally, land cover accounts should be developed over a long period of time, though not necessarily every year. A suggestion would be to develop land cover accounts from 1990 onwards with data compiled every five years. It may also be useful to put together accounts for years further in the past – say 1970 and if possible 1950 such that a longer history can be conveyed.

Land use and land cover accounting are similar in that they provide a broad sense of the changing use of land in a country over time. However, land use accounting focuses on creating a basic link between the land and economic activity. In many cases there will be overlaps between land cover and land use classifications, but some distinctions can be drawn. For example, a given area of forest (land cover classification) may be split into an area that is used for forestry and an area that is protected as a national park (and land use classification).

Land use maps may be found in planning departments where decisions are made regarding the zoning and re-zoning of land. Agricultural censuses and related data sources may also be a source of useful information. There may be existing classification of land use that can be applied. Again, it would be useful to develop a concordance between any selected classification and the international classification proposed in the SEEA. At present the general discussion of land use is focused on terrestrial areas including inland waters, though it is likely to become more relevant in its application to include coastal and possibly marine areas, even out to the edge of the exclusive economic zone (EEZ).

Investigation through this project suggests that Trinidad and Tobago have some history in the development of land-related data but that there is no institutionalised programme to report data on a consistent basis over time. Recently however, there has been an ongoing process to establish the National Spatial Data Infrastructure Programme, which would prepare some core data sets on a regular basis. In this context, the establishment of a regular set of land use and land cover natural capital accounts following international classifications would be of considerable value. The value of these accounts lies not only in providing a picture for the country of broad changes in the composition of land use but can also support the analysis of changes at sub-national administrative levels, the development of maps showing the changes in various locations, and programmes on monitoring economic and environmental performance.

Water

Together with carbon and land cover accounts, water accounting provides the basis for the measurement of the third core indicator for the assessment of ecosystem condition. Further, water's importance in its own right means that the compilation of water accounts can provide information of direct use for policy and analytical purposes.

The methods and techniques for the development of water accounts have been well developed and implemented in many countries. Key aspects of the measurement of water accounts include:

• Tracking the stock of water resources – particularly surface and groundwater – and changes in those stocks through the hydrological cycle (precipitation, evaporation) as well as economic unit/household use through abstraction and return

- Accounting for the abstraction of water from various sources including via desalination and recording the economic units undertaking the abstraction and then understanding its distribution and use through the economy
- Accounting for the collection and treatment of wastewater and its reuse

Ideally, water accounting would also take into consideration water quality, but this aspect has not been fully developed at this stage within the SEEA framework.

As with many small island states, maintaining the supply of, and access to, high quality water is an ongoing challenge for Trinidad and Tobago. While there does not appear to be significant environmental concern at present, there are particular locations, for example in Tobago, where ongoing investigation and investment is needed to ensure that water supply meets water demand.

Natural capital accounting for water represents an important policy support tool by organizing information over time on changes in the available stock of water through precipitation or groundwater and linking this to rates of abstraction. In this context, it is proposed that water accounts be developed for Trinidad and Tobago, and because water accounting tends to be most valuable when implemented for a specific water catchment region, it would be particularly suitable for Tobago. Also, the relevance of water to Tobago's tourism strategies is high, particularly regarding wastewater. The development of such water accounts for Trinidad will likely require discussion on appropriate boundaries for water catchments and watersheds that are meaningful for both data collection and integration into the broader sub-national areas to be used for natural capital accounting.

Biodiversity

Trinidad and Tobago is quite heavily forested with around 60% of the country's area under forest cover. Together with some large wetland and other important ecosystems, Trinidad and Tobago provide important habitats for many species. Recognizing this, the government has established a number of protected areas. Natural capital accounting focusing on the changing condition of those protected areas may be a useful tool to monitor the protected areas and assess the improvements (or otherwise) in biodiversity and related characteristics.

Discussion with wildlife biologists in Trinidad and Tobago indicated that only a limited amount of information on species numbers was available, though data on numbers killed was available for some hunted species which may also be relevant in the management of hunting activity in the country. As part of the extensions to protected areas in Trinidad and Tobago and their improved management, a survey is being conducted on species in those areas and information from such a survey should be able to be used to support compilation of a biodiversity account.

Other aspects of policy relevance of natural capital measurement for Trinidad and Tobago.

Tourism

Tourism is an important economic activity in Trinidad and Tobago, especially Tobago, that underpins many employment and business activities. For both domestic and international tourists, the state of the natural environment is particularly important. This is true in terms of providing attractive and safe areas for recreation, for providing important inputs to production (such as water) and for providing important ecosystem services – such as the flood protection offered by Buccoo Reef – to the seaside tourism facilities. Natural capital accounting can provide useful information that links the economic activity associated with tourism to the environment. Ideally, a natural capital account for tourism would integrate environmental information with data on the economic activity of tourism to support more integrated discussion – for example, linking data on tourism activity in Tobago with data on water availability and supply in the same area.

Food security

With growth in the national economy in recent decades, driven by the oil and gas sector, the relative economic significance of agriculture and food production has diminished. To the extent that there is a desire to address the dependence on imported food products and to support the livelihoods of those living in rural areas, there may be interest in establishing a well-organized set of information on trends in agricultural activity and associated links to the environment. Natural capital accounting may provide this set of information, particularly via the development of specific advice on accounting for agriculture, forestry and fisheries which is currently underway in the FAO (SEEA-AFF for Agriculture, Forestry and Fisheries). Information on environmental aspects including water use, land use, soil quality, fish stocks and the like are as relevant to development and planning decisions as economic information on trade, production and employment.

Factors supporting implementation in Trinidad and Tobago

If natural capital accounting is given sufficient support and a recognition that progress is most likely through taking a step-by-step approach, a programme of accounting could be developed in relatively quick fashion, certainly within three years. From here, it would be expected that a programme could build and improve its outputs over time. The following section considers the critical elements required for successful implementation.

*Recognition of the important role of information
and statistics within decision making*

Without recognition of the benefits that well-compiled and broadly disseminated information can play in the decision-making process, successful implementation of natural capital accounting will not succeed. Natural capital accounting can play a role in ensuring (1) that broad environmental changes and risks are monitored regularly, (2) that decision making takes a broader range of information into account, and (3) that the outcomes of policy decisions can be assessed, including outcomes that may not have been envisaged. There is also general consensus that corporations should compile information on their activities for both their own benefit and the benefit of investors. During the early stages of implementation of natural capital accounting it is unlikely that a lack of data will be the major barrier to success. Rather, experience from many countries suggests that there is often a wide range of relevant data but commonly its availability is constrained by institutional arrangements within a number of government agencies. Since the purpose of natural capital accounting is to integrate environmental data into mainstream economic and development decision making, the available information needs to be shared openly and transparently.

One way of addressing this issue is to develop the potential of the Central Statistics Office (CSO) to be the leading integrator of data across the government and private sectors. While the CSO may not have direct or significant experience in the development of environmental information, they do have expertise in (1) the management of national level datasets, (2) the use of standards and classifications for the integration of data and (3) the dissemination of data to broad audiences. Supporting the application of these skills is an important aspect of the development of natural capital accounting at the national level.

Technical resources and skills

In establishing a programme of work it is recommended that resources be applied to finding and integrating existing data rather than running additional data collections. Not only is it sound practice for the use and re-use of resources but perhaps even more importantly, it places the emphasis on integration and partnership which is essential in this area of work. Even where existing data is utilised, the process of finding and integrating data is not cost free. Resources are required to establish a team that is the focal point for the accounting work. The size of the team will depend on the number and detail of the accounts initially envisaged and the number of other agencies with whom the team will need to engage.

The relevant skill set needed for forming a team for natural capital accounting includes economics, statistics, national accounting, environmental measurement, sustainability policy, geo-spatial analysis, and dissemination and communication. Importantly, those with skills in one or more areas must be willing to appreciate the insights and contributions from the other disciplines such that integration of the data and the development of best practice can be achieved.

Experience from other countries suggests that the national accounting aspects of natural capital accounting may be the most challenging to learn but are core skills required for the task. Thus, it is advantageous to ensure that those with experience with national accounts are included in the natural capital accounts team, and that training programmes on national accounting take place. Furthermore, the natural capital accounts team may be placed within the established national accounts sector, which would have the additional benefit of improving the standard national accounts estimates through consideration of natural capital.

Data availability

The preparation of natural capital accounts requires data – but exactly how much data are required depends on the number and detail of the accounts to be compiled. For example, for the completion of land accounts only information on the stock and changes in stock of land area by type of use and cover is required.

One example of the type of data that would help to support natural capital accounting are the Annual Reports of the Environmental Management Authority which in 2004 provided a comprehensive assessment of the Northern Range essentially using a natural capital accounting approach. A second example is the 2007 release by the ministry responsible for Planning of the First Compendium of Environmental Statistics. While not applying a natural capital accounting approach per se, it covers an impressive range of themes and presents a significant amount of information of direct relevance. Indeed, the national accounts section of the national statistical office led its compilation and it includes an energy account reflecting flows of energy through the economy.

What is most impressive about both of these examples is that their development was very deliberately conducted in an open and collaborative manner, both of which must underpin natural capital accounting.

Further, in terms of progressing natural capital accounting in Trinidad and Tobago it is not necessary that there be a sole dependence on locally sourced data. Increasingly, global data sets based on modelled remote sensing and other data are being constructed. Such data sets will necessarily need to forgo some level of detail nuances that need to be considered at the country level but, at the same time, they may be able to provide a strong supporting role for natural capital accounting work at the national level. Work is currently under way in the form of a combined UNSD/UNEP/CBD project on ecosystem accounting that, among other things, will provide some advice on the use of global datasets for monitoring changes in ecosystem conditions at the national level and assessing flows of ecosystem services.

Conclusions

This section provided an overall assessment of the potential for natural capital accounting in Trinidad and Tobago. We can take from this four key conclusions.

First, the implementation of natural capital accounting in Trinidad and Tobago is not limited by a lack of data. There is evidence of a wide range of data that would be relevant to the compilation of a number of natural capital accounts and that would be sufficient to kick-start a programme of work in this area. Undoubtedly, there are data gaps, particularly at the more advanced end of the natural capital accounting spectrum, e.g. in terms of measuring regulating and cultural ecosystem services. However, programmes to invest in the development of these data would be best implemented once there was a framework in place to compile land and water accounts which should form the basis for any system of natural capital accounts at the country level.

Second, while there are no resources specifically allocated at present to the compilation of natural capital accounts, there is a range of people with diverse and relevant skills in the country who could contribute to the development of a programme of work. The University of the West Indies in particular represents an important asset in promoting and developing the necessary skill sets to advance natural capital accounting work – both in Trinidad and Tobago and in the broader region.

Third, there seems a reasonable degree of high-level ministerial and departmental support for work in this area indicated through the support for this programme of work on ProEcoServ and in various policy documents. Further, the types of policy issues facing Trinidad and Tobago appear to be ones that would benefit from information compiled through natural capital accounting. These issues include, for example, assessing trade-offs in land use, the potential for expansion in the agricultural sector, assessing the tourism sector in Tobago, management of protected areas and hunting, long-term assessments of the oil and gas sector and the management of specific water catchments. In all of these cases, data organized via a natural capital accounting approach would likely prove useful in the decision-making process.

Fourth, while the previous conclusions would indicate a sound basis for establishing an ongoing programme of work on natural capital accounting, there is no clear mechanism within the government of Trinidad and Tobago that would support cross-agency exchange of information and there is no recent evidence of the publication of environmentally related, natural capital information by government agencies in recent years. General support for the national statistical office also appears sub-optimal from a natural capital accounting perspective. Overall, for natural capital accounting to be successful, it will be imperative that an effective cross-agency is established and that there is application of the United Nations Fundamental Principles of Official Statistics.

Piloting a payment for ecosystem services scheme

Ecosystems play a crucial role in providing a wide range of services to society (e.g. flow control in rivers and streams, and clean water to productive soil and carbon sequestration). Individuals, firms, and societies rely on these services for raw material inputs, production processes, and climate stability. At present, however, many of these ecosystem services are either undervalued or not financially valued at all, which leads to their degradation. As part of ProEcoServ-T&T's efforts for mainstreaming ecosystem services in development planning, the objective of this initiative was to provide assistance to the Trinidad and Tobago government in developing a pilot payment for ecosystem services (PES) scheme, which would work towards attaining a number of sustainable development goals – see Box 7.1.

Box 7.1 What is payment for ecosystem services?

Payment for ecosystem services (PES) is a *voluntary transaction in which a well-defined environmental service (ES), or a form of land use likely to secure that service is bought by at least one ESs* **buyer** *from a minimum of one ESs* **provider** *if and only if the provider continues to supply that service* (Wunder, 2005).[7]

Types of PES can include:

1 Public payment schemes for private land owners to maintain or enhance ecosystem services.
2 Formal markets with open trading between buyers and sellers.
3 Self-organized private deals in which individual beneficiaries of ESs contract directly with providers of those services.

Determining the type of payment for the provision of ESs can be based on the following principles

1 Direct investments and opportunity cost which include the investments incurred by the landholders in order to protect the ecosystem that provides the ESs. This approach also takes into account the opportunity costs for not developing other economic activities on the land in order to avoid ESs degradation.
2 Beneficiaries' gains when a beneficiary of the ecosystem services compensates the ecosystem provider with an amount that reflects a market price.
3 Cost of restoring (remediating) an ecosystem affected by a human activity.
4 Economic value of ESs.

Source: For more information: Forest Trends, Katoomba Group & UNEP (2008) *Payments for Ecosystem Services Getting Started: A Primer*: www.unep.org/publi-cations/search/pub_details_s.asp?ID=3996

The most substantial impetus behind implementing PES is that the re-defining and re-structuring of the transaction model causes the benefit to occur where it would not have occurred otherwise. In this context, ProEcoServ has proposed a pilot PES in the Caura Valley in the Northern Range, and in the following sections will offer a road map and guidelines for its implementation. This in turn can help facilitate the achievement of many of Trinidad and Tobago's environmental and development policy goals. For Trinidad and Tobago, a PES scheme is particularly crucial in addressing issues of ecosystem rehabilitation, conservation, environmental education and public awareness, the fragility of the islands' ecosystems,

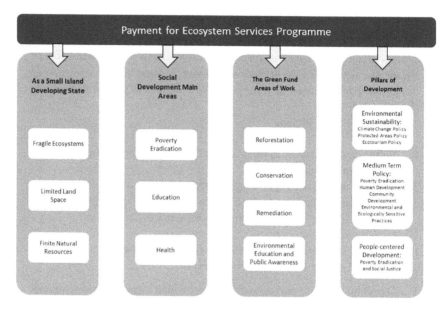

Figure 7.2 Development and environmental policies linked to the PES programme proposal for Trinidad and Tobago.

social goals, as well as climate change policy, protected areas policy and eco-tourism policy, among others. These connections are illustrated in Figure 7.2.

Actors involved in PES pilot in Caura Valley

ProEcoServ conducted a scoping study in the Caura Valley to determine how a proposed PES scheme would be carried out. Caura Valley, with a mosaic of forests, water streams, and agricultural activity, presents a solid opportunity to analyse how the PES scheme would function in a local context in Trinidad and Tobago (Box 7.2) The three main actors involved in the PES scheme for this pilot would be the landholders – providing the ecosystem services (ESs), the local organisation – liaising between the buyer and seller of ESs, and the Green Fund (GF), the government's public environment fund which would be effectively 'buying' the ESs. The actors involved in the proposed structure are shown in Figure 7.3. In this basic structure, several opportunities and potential benefits arise for each of the three actors involved. For example, there is potential to reduce transaction costs, to increase participation, coordination and space for the parties involved, so as to maximize their interests and potential outcomes. This structure has the additional benefits of taking advantage of an existing administrative platform, financial resources, and technical knowledge, for the execution of a PES programme which would fit with local community needs and aspirations.

Box 7.2 The Caura Valley, Northern Range

Located in the Tunapuna-Piarco region of Trinidad and Tobago, the Caura Valley (Latitude: 10° 41' 60 N, Longitude: 61° 21' 0 W) comprises an estimated 4,836 ha representing 4% of the total Northern Range area. With the Tacarigua River originating in the Northern Range and draining into Caura Valley, the area is not only one of the largest watershed areas in the country but most importantly, it is responsible for the provision of drinking water to the country's capital, Port of Spain.

The population is approximately 776 people; the average age range is between 19 and 24. This age group represents an important economic segment of the population with high demands for educational and employment opportunities. Most settlements and farming activities occur within the lower areas of the valley, to the extent that the soil has been classified as of poorer quality in the lower watersheds than in the upper watershed. The majority of farm activities in the Caura Valley take place in small farms averaging five acres with farmers growing crops for the local wholesale market or contract buyers. These farmers cultivate primarily in the rainy season using limited soil conservation technologies such as planting along the contour.

Farming practices also create conditions of vulnerability. The current farming practices by the majority of farmers is the use of harmful non-organic based pesticides which contribute to pollution of water courses, reduction of bio-diversity and the degradation of soils (both in physical structure and soil fertility).

Source: Celestain, Beaumont (2010). *Baseline Assessment of Caura/Tacarigua and Maracas/St. Joseph Watershed.* Final Report submitted to the Cropper Foundation.

Figure 7.3 Tripartite PES Programme Structure for Trinidad and Tobago.

Box 7.3 The Green Fund

The Green Fund (GF) is characterized by a 0.1% (increased to 0.3% in 2016) Green Fund Levy that is applied to gross sales or receipts of a company carrying on business in Trinidad and Tobago. By January 2012, GF had already accumulated funds of USD425 million, making it a powerful funding mechanism for initiatives in remediation, reforestation or conservation of the environment. The creation of the GF was done in the context of a number of policy frameworks that seek to incorporate environmental sustainability in the pursuit of development. Thus, as a grant facility, the GF can function as a catalyst for transforming policies into action. With the involvement of multiple stakeholders – the Ministry of Planning and Development, Ministry of Finance, private sector, civil society organisations – the GF can enable the building of partnerships for greater benefits.

The research and scoping study indicated that a publicly funded PES programme scheme would have the potential to work well in Trinidad and Tobago where the scheme will be voluntary for the sellers (the landholders) but compulsory for society as a primary buyer (through the Green Fund). On one hand, the GF is the representative of society as the administering agency that collects the mandatory 0.1% levy (increased to 0.3% in 2016) on economic transactions in the country. This is the contribution for ESs provisioning from individuals in society through the imposed charge. On the other hand, the landholders who voluntarily decide to participate and those who get selected in the PES programme become the ES's sellers (see Box 7.1).

The landholder as ecosystem provider

It was important to facilitate the enrolment of the population during the application process of PES in T&T, particularly to identify the individuals that own land that provides ESs whose characteristics match the PES policy outcome. Such individuals become the potential sellers of ESs and those who apply to participate in the PES programme will be the pool of candidates to receive compensation. This was achieved through conducting a survey of potential PES participants to attain information on the landholder profile and other socio-economic characteristics.

If the PES intervention is to be structured around environmental issues, with priority placed on favouring vulnerable groups of the population, the design of the programme needs to consider decreasing the transaction costs of the application process for individuals with economic needs. This would allow for other economic activities that do not affect the ESs provision (e.g. ecotourism, agro-ecotourism, sustainable agriculture) to be conducted in the areas incorporated in the PES programme and give priority to small landholders.

The types of ESs that would be provided by the landholders are services directly related to existing ecosystems and their conservation. Such services include provisioning services (water for human consumption and production, timber), regulating services (climate, water flow regulation), cultural services (sense of place for people, aesthetics for visitors, religious), and supporting services (nutrient, water, soils and CO_2 cycles). However, there is room to incentivise actions for remediation of ecosystems, such as reforestation, and environmental education and public awareness activities.

The Green Fund: government's key role for sustained local impact

In the proposed PES scheme for Caura Valley, the national-scale Green Fund was identified as one of the three stakeholders, aside from local landholders and local organisations. The Green Fund is the national environmental initiative that provides funding to community groups and organizations that are engaged in activities related to remediation, reforestation, conservation, environmental education and public awareness. The fund's institutional framework makes it a logical choice of candidate for implementing a successful project, which can as then be a model for a countrywide programme. The PES scheme adheres to a number of policy areas and outcomes that are integral to the Green Fund as a grant facility, fostered through the PES's schemes role in environmental education and public awareness, reforestation, conservation and remediation.

Any PES scheme includes providers and beneficiaries of the ESs. The providers are individuals, communities, and landholders (including the state) that are in the position to protect and secure the provision of the services. A provider becomes a seller when there is a transaction (contractual and economic) for the provision of the ESs based on some known cost. Likewise, the beneficiaries of an ESs are the individuals, communities and landholders in general that receive the ESs as a benefit. However, it is not uncommon that ESs are treated as public goods, free to society, and that many human activities cause damages to the ecosystems that generated the ESs. In relation to the formulation of any PES scheme, addressing this requires considering the principles and complexities of externalities. These principles are directly related to the role buyers, sellers, the beneficiaries, and the providers play in the PES scheme. In many cases resolving an externality problem requires government intervention, particularly when dealing with public goods. In this context, the government in Trinidad and Tobago can play a critical role in rectifying such market failures, which lead to ecosystem degradation. The GF will have to analyse the feasibility and decide on:

- The creation of the PES Sub-Unit within the GF with approval from the Green Fund Advisory Committee (GFAC).
- The type of ESs and PES modalities that will be included as part of the programme.

- A selection of the explicit ESs that can be incorporated by the local organizations in the PES modalities to be implemented at their respective community. This may vary across communities according to local interest. Selecting explicit types of ESs that can be included in the programme that qualify to receive compensation for their provision would help the GF to narrow its priority focus.

In addition to administering the programme, the GF will function as the national funding agency that provides the resources to pay for the ESs provided (the buyer), thus, it is not necessary to create a new mechanism to finance the programme. As Wunder et al. (2008)[8] suggest, in a PES the negotiation process with many sellers (the landholders) is complex and expensive. This responsibility is left to the local organization as intermediary between the GF and the landholders. It is important to identify an entity that serves as intermediary between the sellers and buyers to minimise transaction costs, to facilitate the negotiation process, and to conduct the contractual arrangements.

As administrator, the Green Fund already has a structure that can support a national PES programme through its Executive Unit and the Community Liaison. However, there must be a clear understanding about the role of the Executive Unit in relation to the local organization. At the most fundamental level, the unit should facilitate stakeholder engagement and conduct training on issues around:

- Governance and transparency
- Finance and project management
- Communication and advertising
- Scenario building
- Investment plans
- Monitoring
- Assessment of implementation (verification)
- Reporting
- Best agricultural practices
- Reforestation
- Local tour guides and group management (if applicable)
- Application processes
- Selection of landholders
- Legal concepts and liability

One key element for the success of the programme is the creation of a PES Sub-Unit within the GF. The approval by the Green Fund Advisory Committee (GFAC) for the creation of the PES Sub-Unit is one of the first steps for institutionalizing the PES programme which will be attached to the existing platform of the GF. The proposed ad hoc Sub-Unit will be part of the Executing Unit of the GF but explicitly dedicated to running the PES programme and all that is required for its success. The general functions and responsibilities of the PES Sub-Unit would roughly follow that of the Executing Unit as prescribed by the legislation pertaining to the GF but its function relates specifically to the affairs of the PES programme.

Measuring the effectiveness of any intervention and its impact on policy outcomes (e.g. human welfare, conservation or environmental improvement) entails impact evaluation[9] of the policy using rigorous methodologies. It is important for the GF to conduct impact evaluation studies, with the results utilised in making the necessary adjustments to the interventions and to identify gaps in the proposed PES scheme.

The local organisation's role: liaising between the national and local stakeholders

The local organisation as a liaison body between the Green Fund and the local landholders or ESs providers will be responsible for various aspects of the PES scheme. This includes, but is not limited to:

- Gathering all necessary information about the community – defining local environmental and development priorities
- Providing key biophysical descriptions of the community and its ecosystems
- Selecting and describing the ESs to be provided
- Providing socioeconomic indicators of the community and the landholders
- Profiling the PES participants
- Developing the package with the PES modalities
- Identifying a priori the participant selection criteria

With regards to collaborative work with the GF, the local organisation will be responsible for monitoring, verifying, and reporting implementation activities to all the stakeholders involved, disbursing funds on behalf of the GF, reporting any contracts that need to be reassessed, and other administrative and management activities.

In addition, the local organisation could request to the GF additional funding to finance other activities that support the general GF mandate (e.g. recycling, eco-tourism, environmental education and public awareness campaigns, water quality improvement, etc.).

Knowledge of the local habits, activities, and particularities is crucial for a well-implemented PES programme. For example, research conducted on the Caura Valley indicated that the members of the Community Council had previously been exposed to project interventions which meant that they had prior knowledge and greater receptivity to such eco-finance mechanisms. While local capacity is increasing, the community still needs assistance for the implementation of the PES pilot project.

In this context, it would be strategically apt to approach the Trinidad and Tobago Association of Village and Community Councils (the TTAVCC) to promote the PES programme. The TTAVCC is nationwide community development organization with membership of 500 Village and Community Councils. This would strengthen the participatory decision-making processes and empower stakeholders at the local level.

Legal context of the PES operation

The PES transaction cannot be directly made between the GF and the landholders. The GF legal provisions require that such transaction is made by registered local organisations. In the proposed scheme the local organization works as the intermediary between the GF and the landholders. In any event, the process is formalized with a legal contractual transaction signed by the three parties (i.e. the GF, the local organization and the landholder).

The GF will provide the funds for PES to the local organization according to established GF requirements. In addition, the GF will provide a payment to the local organization as a special provision for managing the PES programme at the local level. This amount will be calculated as a percentage of what is to be paid to each landholder for the duration of active contracts. The disbursement to the local organization will follow the payment schedule of the landholder for whom there is a signed legal contract.

It is fundamental to consider the legal implications in the design and implementing of a PES scheme. This is important to identify whether the PES scheme is in alignment with the legal order of the country. Some of the legal issues linked to the PES scheme include: (1) the land tenure systems, (2) property rights, (3) management of public funds, (4) contractual law, (5) liability and accountability and (6) coupling with other environmental policies. A clear understanding of these issues facilitates pathways for adapting the PES scheme so that it complies with the local or national legal framework.

Payment in PES scheme

The types of compensation provided for the provision of the ESs the PES programme in Trinidad and Tobago would at least include direct monetary payments. Based on the experience from other countries, other forms of compensation could include tax breaks on land property, land rights and land tenure protection, protection by the state from land squatters, technical assistance and other forms of in-kind goods and services and public recognition for contributing to the country's environmental and development policies.

In the case of monetary compensation, the payment should be scheduled in percentage disbursements subject to a series of requirements or conditionality. In addition, in situ verification by the GF and the local organisation should also be scheduled. The specific amount of the compensation per land area per given contractual period for each landholder is not included in this document. This amount needs to be determined based on valuation studies and methods to calculate the ESs cost per area per contractual time. The concept of opportunity cost is commonly used as a guide for determining the compensation amount for the ESs provision. Certainly, this amount varies according to the landholder profile, economic activity, location, etc. As a result, the amount calculated is based on the average net benefit forgone by the ESs provider for refraining from developing a conventional economic activity on the land.

While designing the PES programme for Trinidad and Tobago, it is highly recommended that through prior research and due diligence, a particular priority focus is highlighted for a given policy outcome. The proposed strategy presented advocates for a PES programme that is focused on environmental sustainability through land and natural resource management. By doing so, it aims to facilitate the participation of individuals that are most vulnerable to poverty issues as the primary target population but does not exclude others whose participation would increase the policy impact of implementing the PES programme.

A step-by-step plan for implementation of PES

Implementing the programme by outlining steps, each with a clear objective of what is to be tested, reduces the risk of failure and minimizes the possibility of creating unintended consequences with the policy intervention. For Caura Valley, the following steps are proposed:

Step 1: creation of an updated baseline scenario
of the current situation

Creation of a landholder profile that defines his/her characteristics, land characteristics, the most convenient and feasible forms of compensation the PES programme will offer, existing knowledge about ESs, assessment of willingness to enter a PES programme, preferences in scheduling for payments, and percentage for disbursements and plans for investment by the landholders.

Step 2: information and promotion of the PES programme

This step involves activities to promote the programme through the appropriate communication channels to reach the largest population as possible. The goal here is to drive motivation for participation and addresses possible concerns that may arise from the communities and the landholders. The channels of promotion could be radio, television, communal meetings, newspapers, and others that can be identified based on the landholder profile.

Step 3: identification of criteria for the selection
and application process

The identification of criteria for the selection of landholders or communities for the implementation of the PES should be done a priori based on the GF priorities and the landholder profile. All the applications that comply with the requirements that are submitted during the time of application have to be accepted. It is likely there will be more applications than the GF can fund. Thus, a selection process with clear, defined rules needs to be developed prior, to avoid misunderstandings and social conflicts. To facilitate the participation of the most vulnerable groups, promotional activities should be organised in the communities to offer guidance during the application process. Ideally, applications can be submitted at locations where the landholder lives to avoid having them make long trips to urban centres. Avoid asking for unnecessary requirements – for example if proof of land ownership is important, find the basic means that most feasibly serve the purpose (i.e. request property title but also allow for alternatives such as leasing documentation, letters of comfort, letters signed by owners of properties around, etc.). Additionally, an application fee waiver in Phases 1 and 2 may increase participation of the poor.

Step 4: selection of participants for the PES programme

The use of the selection criteria helps identify the priority areas that are ecologically fragile or at risk of deterioration (degradation or deforestation). Priority will be given to landholders located in such areas starting with applicants that are economically more vulnerable. Those environmental and socio-economic priorities can vary their hierarchies from region to region within Trinidad and Tobago, but the selection rules and criteria need to be clearly defined beforehand. Criteria for selection may be based on GF's areas of work, including remediation, reforestation, conservation, and environmental education and public awareness. Other elements for selection include land uses and socio-environmental priorities of the community. These criteria form the basis for measuring change from the baseline compared to the outcomes post-policy intervention. Once participants have been selected, contracts are to be signed.

Step 5: implementation for the PES programme

The PES contract outlines the responsibilities between the parties according to the PES modalities. The schedule of payment and the percentage are explicitly stated in the contract as well as other general provisions.

Step 6: monitoring, evaluation, reporting and adjustments

Monitoring will be conducted by both the GF and the local organization; the GF can also conduct independent monitoring activities. After the evaluation assessment, a report from the local organization to the GF needs to explicitly state whether the percentage payment that corresponds should be executed or not. The report can recommend if a contract needs to be temporarily suspended or terminated. The monitoring and evaluation process will also serve to make adjustments to the programme when considered necessary to guarantee the success of the PES programme.

Way forward: PES in phases

Implementing the proposed PES programme in three broad phases is the strategy most likely to be successful. Within each phase of implementing the proposed PES programme, a series of timeline steps (as indicated in the previous section) in each of the three phases would lead to the completion of each phase. Each step represents one level of complexity in the design.

Phase 1 consists of a two-year PES pilot project in the Caura Valley community. The goal of the pilot is to develop experience and capacity as well as to collect information to inform the other phases of the programme. **Phase 2** will start at the end of the second year of the pilot project and consists of the implementation of the PES programme at the national level. **Phase 3** starts at the end of the third year of phase 2 (five years after the launch of the pilot project). This is the consolidation phase and several adjustments will need to be made in the overall design. For each phase the GF needs to publish in a PES manual the specific requirements, guidelines, rules, deadlines and any operational issue regarding the application and implementation steps.

Scaling-up of the programme in phases provides opportunities to improve the design with new information for decision making. It is also a way to monitor progress of how communities and landholders respond to the implementation of the PES scheme. As the programme consolidates, the perception about ESs by landholders changes from a situation without the policy intervention. In this case, the benefits that society gets from ESs are not fully recognized as such and landholders are simply providers not aware of their contribution; likewise, landholders may not be aware that their actions on the land can also cause ESs degradation. With the intervention, the landholders become more aware of the value of ESs to society and their contribution to the provision of such benefits are recognized by society.

Overall, while this report provides general guidelines for the design of a PES scheme in Trinidad and Tobago, the implementation of PES requires special attention on the creation of a PES programme that will facilitate realizing many of the country's environmental and development policy goals. A PES scheme can be designed in a way that is aligned with other policies that address issues around ecosystem rehabilitation, conservation, and social goals. In the context of Trinidad and Tobago, the national-level Green Fund is an important catalyst for mainstreaming initiatives such as PES and subsequently achieving environmental and socio-economic goals at different scales. PES is a concept that must be contextualized and shaped according to local particularities. The achievement of goals or desired outcomes should be assessed and evaluated at every stage to understand overall success. Finally, PES schemes rely on local interest, dissemination of appropriate information and knowledge, access to expertise or technical knowledge and assistance, compensation, and a steady monitoring and evaluation to assess the impact and success of implementation.

Notes

1 The valuation of downstream or off-site effects of soil erosion proves to be particularly difficult due to the wide variety of off-site effects and the difficulty of quantifying a mathematical relationship between erosion and downstream impacts such as flooding and ecological disturbance.
2 There is also increasing documentation of the value of soil diversity in maintaining belowground biodiversity, which is critical to the functioning of aboveground ecosystems. See: Bardgett, R. and W. van der Putten. 2014. Belowground biodiversity and ecosystem functioning. *Nature* 515, 505–511 (27 November). However, this value is not addressed here.
3 Intergovernmental Panel for Climate Change (2007). *Climate Change 2007: Impacts, Adaptation and Vulnerability*, Cambridge: Cambridge University Press.
4 Mumby, P., Edwards, A., Arias-Gonzalez, E., Blackwell, P., Gali, A., Gorczynska, M., et al. (2004). Caribbean, Mangroves enhance the biomass of coral reef communities in the. *Nature*, 533–536.
5 World Bank, Trinidad and Tobago Country Profile. http://data.worldbank.org/country/trinidad-and-tobago.
6 For a complete example of option 4, see the UNEP publication *Guidance Manual on Valuation and Accounting of Ecosystem Services for Small Island Developing States*. www.esevaluation.org/images/Guidance%20Manual%20SIDS%20Full%20Report.pdf.
7 Wunder, S. (2005). Payments for Environmental Services: Some Nuts and Bolts. CIFOR Occasional Paper No. 42. Center for International Forestry Research, Bogor, Indonesia.
8 Wunder, S., Engel, S., and Pagiola, S. (2008). Taking stock: A comparative analysis of payments for environmental services programs in developed and developing countries. *Ecological Economics*, 65(4), 834–852. http://dx.doi.org/10.1016/j.ecolecon.2008.03.010
9 Baylis, K., Honey-Rosés, J., Börner, J., Corbera, E., Ezzine-de-Blas, D., Ferraro, P. J., . . . Wunder, S. (2015). Mainstreaming impact evaluation in nature conservation. *Conservation Letters*, 9(1): 58–64 n/a-n/a. doi:10.1111/conl.12180. Ferraro, P. J. (2009). Counterfactual thinking and impact evaluation in environmental policy. *New Directions for Evaluation*, 2009(122), 75–84. doi:10.1002/ev.297. Gertler, P. J., Martinez, S., Premand, P., Rawlings, L. B., and Vermeersch, C. M. (2011). *Impact Evaluation in Practice*. World Bank Publications.

References

Bardgett, R. and van der Putten, W. (2014). Below ground biodiversity and ecosystem functioning. *Nature* 515, 505–511 (27 November).
Baylis, K., Honey-Rosés, J., Börner, J., Corbera, E., Ezzine-de-Blas, D., Ferraro, P. J., and Wunder, S. (2015). Mainstreaming impact evaluation in nature conservation. *Conservation Letters*. doi:10.1111/conl.12180.
Celestain, B. (2010). Baseline Assessment of Caura/Tacarigua and Maracas/St. Joseph Watershed. *The Cropper Foundation*. 48 pg. https://docplayer.net/21972086-Baseline-assessment-of-caura-tacarigua-and-maracas-st-joseph-watersheds-final-report-submitted-to-the-cropper-foundation.html
Dempewolf, L. (2018). Identification, Assessment and Valuation of Pollination Services in Neotropical Agricultural Landscapes, Trinidad W.I. Ph.D. thesis, University of the West Indies, St Augustine. 307 pg.
Ferraro, P. J. (2009). Counterfactual thinking and impact evaluation in environmental policy. *New Directions for Evaluation*, 122, 75–84. doi:10.1002/ev.297
Food and Agricultural Organization of the United Nations and the United Nations (2015) SEEA for Agriculture, Forestry and Fisheries, Global Consultation Draft, December, FAO, Rome.

Forest Trends, Katoomba Group & UNEP (2008) *Payments for Ecosystem Services Getting Started: A Primer*. www.unep.org/publications/search/pub_details_s.asp?ID=3996

Gertler, P. J., Martinez, S., Premand, P., Rawlings, L. B., and Vermeersch, C. M. (2011) *Impact Evaluation in Practice*. World Bank Publications. http://documents.worldbank.org/curated/en/698441474029568469/Impact-evaluation-in-practice.

Intergovernmental Panel for Climate Change (2007) *Climate Change 2007: Impacts, Adaptation and Vulnerability*. Cambridge: Cambridge University Press.

Mumby, P., Edwards, A., Arias-Gonzalez, E., Blackwell, P., Gali, A., Gorczynska, M., et al. (2004). Caribbean, Mangroves enhance the biomass of coral reef communities in the. *Nature*, 533–536.

Rawlins, M. (2017). Ecosystem Services Approach for Natural Resource Planning and Management in Trinidad and Tobago. Ph.D. thesis, The University of the West Indies, St Augustine. 342 pg.

UNEP (2014). *Guidance Manual on Valuation and Accounting of Ecosystem Services for Small Island Developing States*. Nairobi, Kenya: UNON Publishing Service, 112 pg.

United Nations (2013). *SEEA Implementation Guide*. New York: United Nations Statistics Division.

United Nations, European Commission, Food and Agricultural Organization of the United Nations, International Monetary Fund, Organisation for Economic Co-Operation and Development, The World Bank (2014a). *System of Environmental-Economic Accounting 2012: Central Framework*. New York: United Nations.

United Nations, European Commission, Food and Agricultural Organization of the United Nations, Organisation for Economic Co-Operation and Development, The World Bank (2014b). *System of Environmental-Economic Accounting 2012: Experimental Ecosystem Accounting*. New York: United Nations.

World Bank, Trinidad and Tobago Country Profile (2019). http://data.worldbank.org/country/trinidad-and-tobago.

Wunder, S. (2005) *Payments for Environmental Services: Some Nuts and Bolts*. CIFOR Occasional Paper No. 42. Center for International Forestry Research, Bogor, Indonesia.

Wunder, S., Engel, S., and Pagiola, S. (2008). Taking stock: A comparative analysis of payments for environmental services programs in developed and developing countries. *Ecological Economics*, 65(4), 834–852. doi:10.1016/j.ecolecon.2008.03.010

Yoder, D. C., G. R. Foster, G. A. Weesies, K. G. Renard, D. K. McCool, and J. B. Lown. (2001, updated July 2004). Evaluation of the RUSLE soil erosion model. In: Parsons, J. E., Thomas, D. L., Huffman, R. L. (eds.). *Agricultural Non-Point Source Water Quality Models: Their Use & Application*, 107–116, Southern Cooperative Series Bulletin #398. http://citeseerx.ist.psu.edu/viewdoc/download?doi=10.1.1.621.3389&rep=rep1&type=pdf#page=116

8 Nature's services facilitate National Green Growth Strategy
Vietnam 2025

Nguyen Van Tai, Kim Thi Thuy Ngoc,
Michael Parsons, Jana Juhrbandt,
Tran Thi Thu Ha, Tran Trung Kien,
Gregg Michael Verutes, Ngo Chi Hung and
Le Thi Le Quyen

Introduction

The work carried out in ProEcoServ-Vietnam (ProEcoServ-VT) was a collaborative effort led by the Institute of Strategy and Policy on Natural Resources and Environment (ISPONRE) in partnership with the Ministry of Natural Resources and Environment. The main objective of ProEcoServ-VT is to support the Ca Mau Department of Natural Resources and Environment and the Ca Mau National Park management to integrate ecosystem services (ESs) into land use planning. For this reason, a pilot valuation study was implemented. The pilot work took place on the Cape Ca Mau National Park, in the Ca Mau Province, that covers 12% of the country and comprising one of the largest remaining contiguous mangrove forests in Vietnam (Box 8.1).

Box 8.1 Cape Ca Mau National Park

Ca Mau is the southernmost province of Vietnam, with the total area of the province estimated to be 529,487 ha including 462,968 ha of agricultural land accounting for 87.44% of the total area; 57,974 ha of non-agricultural land occupies 10.95% of the total area; while 8,545 ha (1.61%) is unused land. The total forest area of the province is 99,173 ha with mangrove-type forests comprising the largest area with 62,436 ha. The Cape Ca Mau National Park (also known as Mui Ca Mau) is located within the Ngoc Hien District of the province. The Mui Ca Mau houses a rich mangrove ecosystem, among other ecosystems and natural features. In fact, the Ngoc Hien District has the largest area of mangrove forests in the province with

43,523 ha attributed to its long stretch of coastal areas. Ca Mau's mangrove area was listed in the Recommendation of National Marine Priority by the Ministry of Aquiculture in 2005. It was also nominated as a coastal Ramsar site in 2006. It was also recognized as the 5th Ramsar of Vietnam and the 2,088th of the world in 2012. More recently, the Vietnamese prime minister approved the National Action Plan for Biodiversity to 2010 and Strategy to Implement Biodiversity Convention to 2020, within which Mui Ca Mau is one of the critical sites for conservation of biodiversity in Vietnam – being today a UNESCO Biosphere Reserve and a Ramsar site. Today the park is managed by using National Policies for National Protected Areas and Forests by Decision No 142/2003/QD-TTg and Decision 08/2001/QD-TTg of Prime Minister, for Protected Mangrove Forests.

Source: ProEcoServ-VT; UNESCO

Banking on the results of this study, ecosystem services are now present in the land-use and planning policy-making processes, at both the regional and the national level. In fact, the ProEcoServ-VT team took the lead in making Increasing Investments in Natural Capital in the Greater Mekong Sub-region as the theme of the Fourth Greater Mekong Sub-region Environment Ministers' Meeting held in 2015. One of the objectives of the high-level session was mainstreaming natural capital considerations into socio-economic planning and investment decision making processes. The Vietnamese officials presented ProEcoServ-VT as a "best practice" initiative in the mainstreaming session of this meeting.

Setting the scene

The activities of ProEcoServ-VT aimed to inform decision makers of the wide range of economic values that mangrove ecosystems provide to the population of Vietnam. Since most of these values are not directly associated with a market price, their economic value is often overlooked or underestimated by policy makers and land use planners. To equate non-priced benefits as with zero economic value is misleading and incorrect. Therefore, the main strategy of this project was to develop and implement a pilot economic valuation study of the mangrove forests in Ngoc Hien, the southernmost rural district of Ca Mau Province. In this context, the project proposed involved (1) developing valuation tools that capture the true economic value of mangroves, with the results made available to land use planning policy in the Cau Mau Province; (2) raising awareness and capacity building activities for national and provincial decision makers on ESs and (3) mainstreaming the uptake of ESs in the policy agenda in Vietnam, at both provincial and national levels.

Economic value of mangrove ecosystem services

Setting the scene and the rationale for the valuation exercise

Mangroves in Ca Mau Province support millions of people with market priced goods, including wood, timber and food. At the same time, this ecosystem provides a wide range of services to the local communities, whose benefits are often not as tangible or measurable market prices, such as flood control, water purification and erosion reduction and the provision of habitat that supports rich biodiversity of hundreds of fauna and flora species, some of which are endangered. Indeed, BirdLife International listed 4,388 ha in Ca Mau as an Important Bird Area (IBA), and in 2009 it was established as a UNESCO Biosphere Reserve. It was also recognized as the 5th Ramsar of Viet Nam and the 2,088th of the world in 2012. Thus, the main objective of this project was to implement a pilot valuation study of the mangroves in Ca Mau Province in order to assess the true economic value of mangrove ecosystems, including both market and non-market priced benefits. This information can then be integrated into spatial planning and development policies for the Ca Mau Province, by way of providing a significant contribution to the National Green Growth Strategy.

Ca Mau's mangrove forest area, notably located in the Ngoc Hien District, is about 43,523 ha, including 18,762 ha of production forests, 12,765 ha of protected forests and 11,996 ha of special-use forests. Currently, most of the mangrove forests have been allocated to forestry companies, forest management committees and military units and only around 5% of the total mangroves (production forests) are allocated to households. The exploitation and utilization of mangroves in Ngoc Hien is fully compliant with the guidelines of the Ministry of Agriculture and Rural Development as well as documents related to forest management plans for each stage of the Department of Agriculture and Rural Development in Ca Mau. Therefore, for mangroves that are production forests, forest owners are allowed to cut trees in the entire area when the forest reaches the age of exploitation (usually 12–15 years). After exploitation, forest owners have to re-plant the forests for 12 months with funds deducted from the profits of exploitation. For protected forests, when the forest is up to standards for exploitation, forest owners are allowed to exploit them in the form of clearing in each group. For special-use forests, the exploitation can only be conducted in some areas of protected landscape areas and under close monitoring of the administration services. The harvests are mainly fallen trees, dead trees and non-timber forest plants.

It is proposed that a valuation study be implemented that assesses the services provided by mangroves, including timber and firewood, commercial fish nurseries and aquaculture benefits (mangrove provisioning services), coastal protection, carbon sequestration benefits (mangrove regulating services) and recreational benefits (mangrove cultural services). The economic values of these services are presented in the next paragraphs. All economic value estimates are expressed in 2014 prices.

Mangrove provisioning services

Timber is one of the main products provided by mangroves and is often used for construction of houses and fishing boats, among other purposes. It is harvested through forest timbering activities, which are carried out every 12 to 15 years, depending on forest quality. In the most cases, households, following contracts signed with mangrove forest management boards, do the timbering activities. Harvested timber is sold at local markets. Table 8.1 summarizes a number of indicators measured in mangroves used in this study so as to characterize the bio-physical productivity of this asset.

Combining this information with exploitation data of the various types of mangroves and the market prices associated to the different mangrove outputs, as shown in Table 8.2, we can characterize the income derived from forest production of mangroves.

In addition, this economic activity involves production costs, ranging from labour costs with management, planting and cut as well as shipping costs. The production costs depend on factors such as terrain conditions and technical design (seeds, cost of the land, density, planting techniques, care and protection) with an

Table 8.1 Indicators measured in mangroves of exploitation, Ca Mau Province

Indicators	*Unit*	*Production forests*	*Protected forests*
The average height of trees when cut	m	10.50	12.00
The average diameter of trees when cut	cm	10.05	11.00
Density	tree/ha	2,150	2,400
Volume	m³/ha	83.75	105.10

Source: Management Board of Protected Forests in Ca Mau

Table 8.2 Different types of mangrove forest outputs

Output	*Unit*	*Production forests*	*Protected forests*
Timber	m³/ha	8.53	7.48
	US$/m³	34.50	53.90
Firewood (high quality)	stere/ha	69.34	98.38
	US$/stere	31.30	37.70
Firewood (low quality)	stere/ha	14.29	15.19
	US$/stere	24.80	24.80

Note: The stere is a unit of volume in the original metric system that is used for measuring large quantities of firewood or other cut wood and it equals to one cubic metre – see https://en.wikipedia.org/wiki/Stere

Table 8.3 Net income from the exploitation of mangrove forests (per cycle)

	Production forests	Protected forests	Average
Income from timber and firewood US$/ha)	3,120	4,087	3,603
Land-use tax @ 4% (US$/ha)	125	164	145
Production costs (US$/ha)	862	862	862
Net income (US$/ha)	2,133	3,061	2,596

average estimate of US$862/ha. Income is also subject to taxation. The net income from mangrove exploitation is presented Table 8.3.

Thus, currently the net income from timber and firewood for forest owners ranges from 2,133 to US$3,061/ha, per cycle. One cycle is 12–15 years, so the net income from the exploitation of mangrove forests ranges between US$164–235/ha/year.

There are two main types of fishing in Ngoc Hien District: inshore and offshore fishing. Although mangroves affect both types of fishing, the study assessed these impacts on inshore fishing, as it plays a fundamental role in supporting the livelihoods of communities in the Ngoc Hien District. Currently there are 488 households participating in inshore fishing, with a collective annual harvest of about 24,020 tons. The equipment used for inshore fishing is relatively simple technology, involving as trawls, grill nets and squid. Fishery harvest is rich and diverse in species including shrimps, crabs, squids, fish, snails, and clams.

To estimate the value of inshore fisheries, interviews were conducted with 50 representative households who regularly participate in inshore fishing (Table 8.4) A total of 488 households participate in inshore fishing, which amounts to US$2.19 million/year in aquatic resource value, or US$50/ha/year, on average.[1]

To clarify the relationship between mangrove forests and fisheries production, a quantitative research has been conducted using the same approach used for valuation of mangroves for fisheries in Thailand by Barbier and Cox (2004). According to this article, the area of mangroves is an important parameter in the equations explaining the relationship between fisheries production and (1) efforts for exploitation (as shown by the number of vehicles or total extraction time or capacity of vehicles used for exploitation) and (2) the area of mangroves (supply of food and safe shelter for aquatic species). Econometric estimation results show that the impact of these two parameters is statistically significant. In particular, if the mangrove area does not change and the number of households licensed for fishing increased by 10%, the volume of fisheries exploitation will be increased by 3.8%. Otherwise, if the number of households licensed for fishing does not change and the area of mangrove is increased by 10%, the volume of fisheries exploitation will be increased by 1%. Under the assumption that the results from Thailand can be extended to Vietnam, this means that a 10% increase in the mangrove area will be associated with an increase in the annual income of the fishermen in the

Table 8.4 Income from inshore fishing in Ngoc Hien

Species	Number of households interviewed	Average volume (kg/household/year)	Price (US$/kg)	Income (US$/year)
Shrimp	23	500	11.50	132,250
Crab	6	275	8.28	13,662
Squid	4	494	7.13	14,089
Fish	19	520	3.68	36,358
Snail	14	982	1.84	25,296
Clam	4	400	1.38	2,208
Total income of households interviewed (US$/year)				223,864
Average income of households (US$/year)				4,477

mangrove between US$21,849 and 177,957, depending on the final composition of the species harvested. In other words, a 10% increase in the mangrove area will be associated with an increase in annual income between US$0.50 and 4.10/ha from inshore fishing in the Ngoc Hien mangroves.

In 2013, the area for aquaculture of Ngoc Hien was 24,222 ha, of which an area of 10,270 ha was used for ecological cultivation, accounting for over 40% of the total area. Currently, this form of cultivation was only applied to the shrimp species, and the productivity is higher than traditional forms of cultivation. Specifically, the production of ecological cultivation is 650–750 kg/ha/year and that of traditional cultivation is 250–300 kg/ha/year.

The idea of mangrove forests affecting aquaculture production was proposed firstly by Odum and Heald (1975). This idea was strengthened by various studies around the world. These studies are based on three hypotheses to explain the relationship between mangroves and aquaculture production in coastal areas. These are: (1) mangroves provide food for aquatic species, (2) mangroves provide shelter for aquatic species from physical disturbance and predators and (3) mangroves can reduce pollution affecting aquatic species, thus, enhancing the health and viability of aquatic species under the forest canopy. Results of studies on the relationship between mangroves and aquaculture production in a number of regions of the world are summarized in Table 8.5.

To estimate the impact of mangroves on aquaculture production, it is prescribed to collect information related to each cultivation model including: area, depth of the pond, production costs (labour, breeding, food), water quality, mangrove cover, and fishery production. However, due to the limited time and resources, this study uses only general information related to aquaculture production, cultivation area and mangrove forest area to analyse data from 2004–2013. The equation used in this study was as follows:

$$h = AE^a S^b Ln(h_i) = A_0 + aLn(E_i) + bLn(M_i) + \mu_i$$

With:

h_i aquaculture production harvested in year i (ton)
E_i the area of aquaculture in year i(ha)
Mi the area of mangrove forests in year i (ha)

Estimated coefficients are positive and statistically significant. Specifically, if other factors are not changed, the expansion of aquaculture area by 1% will increase aquaculture production up to 2.95%. Whereas, if the area of mangroves increased by 1%, the aquaculture production will be increased by 1.96%.

Table 8.5 Study results on the relationship between mangroves and aquaculture

Study areas	Relationship	Correlation (sample number)	Reference
Tropical regions of the world	Shrimp productivity – vegetation areas	0.54(27)	Turner (1977)
Tropical regions of the new world	Shrimp productivity – vegetation areas	0.64(14)	Turner (1977)
Indonesia	Shrimp productivity – mangrove areas	0.89(N/A)	Martosubroto and Naamin (1977)
Carpentaria-Australia	Shrimp productivity – length of mangrove forests	0.58(6)	Stapbles et al. (1985)
Gulf of Mexico	Fish productivity– mangrove areas	0.48(10)	Yanez-Arancibia et al. (1985)
Tropical regions on the world	Shrimp productivity – mangrove areas	0.53(N/A)	Pauly and Ingles (1986)
Peninsular Malaysia	Shrimp productivity – mangrove areas	0.89(10)	Paw and Chua (1991)
Philippines	Shrimp productivity – mangrove areas	0.61(18)	Paw and Chua (1991)
Philippines	Fish productivity – mangrove areas	0.34(18)	Paw and Chua (1991)
Vietnam	Productivity of fish and shrimp – mangrove areas	0.95(N/A)	de Graaf and Xuan (1998)
Vietnam	Shrimp productivity – mangrove areas	0.88(5)	de Graaf and Xuan (1998)
Tropical regions on the world	Shrimp productivity – mangrove areas	0.38(37)	Lee (2004)
New South Wales-Australia	Fish productivity – total area of mangrove and wetland and sea grass strip	0.32–0.75(49)	Saintilan (2004)
Malaysia	Shrimp productivity – mangrove areas	0.37–0.70 (36)	Loneragan et al. (2005)
Queensland-Australia	Shrimp productivity – mangrove areas	0.37–0.70 (36)	Manson et al. (2005)

Source: Alongi (2009)

Thus, if the area of mangroves increased by 60%, the aquaculture production will increase by 117%. This result is relatively reasonable compared to the reality of the aquaculture industry in the Ngoc Hien District in recent years.

According to the statistics of Ca Mau Agency of Statistics, in 2012, aquaculture production of Ngoc Hien recorded an income of US$87.5 million (Provincial Statistical Yearbook of Ca Mau, 2012). With a total of 51,646 ha allocated to aquaculture production, this corresponds to an income from aquaculture in the Ngoc Hien mangroves of about 1,695 US$/ha/year. Between 2011 and 2012, the change of cultivation area is insignificant. On the other hand, the area of mangroves increased by 5.67% and the aquaculture production increased by 2,545 tons, from 49,010 tons in 2011 to 51,555 tons in 2012. The value of the increased aquaculture production is valued at US$ 4,154,808. Thus, on average, an increase in area of mangrove by one ha is contributing to a US$187/year increase in value of aquaculture production.

Coastal protection, carbon sequestration benefits (mangrove regulating services)

Mangrove forests are well known for their role in coastal protection from storms and other forms of extreme weather events. Recent studies indicate that forest mangroves with a coastal belt of 100 m, or more, are able to reduce the height and power of waves by 50% (Das and Jeffrey (2009). This way, the local communities that are living in these coastal regions suffer less damage than those who are living in coastal regions with destroyed mangroves. In order to estimate the benefits from coastal protection that mangrove forests provide to the coastal community of the Ngoc Hien District, the Expected Damage Function (EDF) is used. According to the economic valuation approach, the economic value of coastal protection refers to the expenditure saved in terms of forgone costs incurred to property, infrastructure and production when mangrove services that protect economically valuable assets are lost.

Two critical steps to implementing this approach include: (1) analysing the influence of mangrove area on the expected incidence of economically damaging natural disaster events and (2) monetizing economic damage incurred per event. From the operative perspective, it is necessary to collect data on the incidence of past storms and changes in mangrove area in coastal regions. Related to extreme weather events, information was collected on impacts on human lives, including dead/missing and injured; damage to human-built assets, including damaged houses, roads and boats; as well as damages on production, including losses of income from agriculture and fishery activities. Table 8.6 summarizes the occurrence and intensity of damage caused by storms and other forms of extreme weather events in Ngoc Hien in recent years.

To estimate the value of protective services on mangrove forests, the study has analysed the relationship between damage from a disaster and the area of mangroves between 2007 and 2013. According to Barbier (2007), a number of factors will affect the intensity of a disaster in a certain period of time, including: (1) the

Table 8.6 Damage caused by natural disasters in Ngoc Hien

	No. of disasters	Dead/ Missing	Injured	Destroyed houses	Destroyed boats	Total damage (in US$)
2007	7	–	2	92	–	217,590
2008	7	–	–	6	–	147,610
2009	8	1	1	82	3	533,048
2010	11	–	–	21	–	776,756
2011	8	5	–	–	4	418,968
2012	7	–	–	19	4	212,152
2013	6	–	–	55	6	117,162

Source: Department of Statistics of Ngoc Hien District

number of events, (2) the area of mangrovesand (3) the socio-economic conditions of localities. The functions describe the relationship between parameters as following:

$$\ln C_t = \alpha + \beta_E E_t + \beta_M M_t + \beta_x X_t + \mu_t$$

With:

C_t the total damage caused by natural disasters in the year t, with $t = 2007$, 2008 . . ., 2013

E_t the number of extreme weather events (storm, tidal) in the year t

M_t area of mangroves (ha) in the year t

X_t socio-economic factor that can affect the level of damage

Results of regression analysis show that the coefficient associated to the number of extreme weather events is positive, and its parameter estimate is 0.137. This means that when other factors are unchanged, any additional extreme weather event increases the socio-economic losses by 13.7%. Similarly, the coefficient of area of mangroves is negative and its parameter estimate is −0.01649. This demonstrated that when other factors are unchanged, an increase in one ha in the area of mangroves will be associated with a decline in 1,649% of the total damages.

According to the statistics for the period of 2007–2013, Ngoc Hien experienced 54 extreme weather events causing damages of US$2,475,674. The average cost of damages for each of the recorded extreme weather events is about US$45,000. With a 1 ha increase in the forest mangrove area the expected damage is reduced by US$742 to 756. In other words, the annual economic value of the protective function of mangrove forests is valued at the range of US$742 to 756/ha.

Mangroves have an important role in terms of as carbon sequestration services. Similar to other ecosystems, mangrove forests have five main types of mechanisms for carbon storage, including: (1) wood tree biomass above the ground, (2) wood tree biomass below the ground, (3) dead trees, (4) litter and (5) soil/sediments. As far as mangroves are concerned, reservoirs in litter usually account for a very small

Table 8.7 The ratio of major carbon stocks in various mangrove forests

Country	Research area	Genus	Above ground (%)	Underground (%)	Soil, in sediment (%)
Micronesia[1]	Coastal area	*Sonneratia*	18	14	68
Palau[1]	Coastal area	*Rhizophora*	16	11	73
Bangladesh[2]	Estuary	*Avicennia*	17	8	75
Indonesia[3]	Estuary	*Rhizophora*	12	5	83
Average			15.75	9.5	74.75

Note: (1) from Kauffman et al. (2011), (2) from Donato et al. (2011), and (3) from Murdiyarso et al. (2010)

proportion. The reservoirs in dead trees also are not large unless entire ecosystems have recently undergone physical disturbances such as natural disasters or land use changes. The reservoirs in wood biomass of mangroves are the most noticeable because they account for a large proportion of the total amount of carbon of ecosystems (Alongi, 2009). Table 8.7 shows the ratio of major carbon stocks in various types of mangrove forests. The reservoirs of carbon in biomass of trees (both above and below ground) account for only a relatively small proportion (average of 25.25%) compared to other reservoirs in soil (average of 74.75%). With regard to mangrove ecosystems of Ngoc Hien District, to estimate the value of carbon storage, the ProEcoServ-VT team determined the amount of carbon in the biomass above the ground (including aboveground roots). For the soil reservoir, due to limited time and human resources, it will be determined based on the equivalent proportion as shown in Table 8.7.

The amount of carbon in the biomass aboveground is assessed by estimating the fresh biomass of parts such as trunk, branch, leaf etc. Secondly, fresh biomass samples are taken to the laboratory to determine dry biomass and the amount of carbon in dry biomass (the weight of carbon usually accounts for 48–52% of the dry biomass weight). Finally, to be converted to equivalent CO_2, it is multiplied by 3.667, as one ton of carbon equals $44/12 = 11/3 = 3.67$ tons of carbon dioxide.

Based on analysis of data of the total amount of carbon stored in sampled trees and the density of reservoirs of carbon in mangrove forests, Table 8.8 shows the estimation results of the amount of carbon in main reservoirs of various types of mangroves in Ngoc Hien District.

Table 8.8 shows that the amount of carbon in biomass (above and below ground) and entire forests (biomass and soil) depends on the age of the tree, on the ability of each tree species to absorb CO_2 and the density of forests. In many cases, forests that are older and with low densities can absorb the equivalent amount of carbon as those are at high densities. Soil biomass is calculated with the proportional principle as shown in Table 8.8, i.e. soil biomass is approximately 4.5 times greater than aboveground biomass.

With regard to mangrove forests in Ngoc Hien, the amount of CO_2 used in the computations is the one related to the tree biomass (above- and belowground)

Table 8.8 Estimated CO_2 content in different mangroves in Ngoc Hien District

	Age	Density (tree/ha)	Relevant CO_2 content (ton/ha)				
			Aboveground biomass (A)	Underground biomass (B)	Tree biomass (A)+(B)	Soil biomass (C)	Forest biomass (A)+(B)+(C)
Rhizopho-	3	7,500	25.142	15.09	40.23	30.85	71.082
raapiculata	7	4,500	62.476	37.49	99.96	76.67	176.631
Blume species	13	4,016	172.285	103.37	275.66	211.42	487.076
	16	3,766	297.326	178.4	475.72	364.87	840.589
	18	3,150	289.757	173.85	463.61	355.58	819.191
	n/a	2,790	107.096	64.26	171.35	131.42	302.777
Avicennia	n/a	2,770	150.57	90.34	240.91	184.77	425.686
alba species	n/a	2,660	207.058	124.24	331.29	254.09	585.388
	n/a	2,540	328.87	197.32	526.19	403.58	929.769
Avicenniaoff-	6	4,800	335.898	201.54	537.44	412.2	949.638
cinalis species	10	3,900	579.258	347.55	926.81	710.84	1637.65

because mangrove forests are usually cleared to increase the area of aquaculture. In that case, the amount of CO_2 emitted is located mainly in the biomass of mangrove trees. There are very few cases where land use changes lead to draining water from forests and soil improvement that causes the loss of the amount of CO_2 in soil. In order to calculate the economic value of the carbon storage services, information is needed on the market price of carbon. In this study, a conservative value of US$8 per ton of CO_2 was used, which is in agreement with the price data provided by the European Emission Allowances–Europe Energy Exchange, a well-known carbon market, and its information is often used in putting a price on carbon. The results are presented in Table 8.9.

Results show that mangrove forests, when managed and protected, provide a significant economic benefit, as measured from the sale of carbon emissions credit, amounting to US$2,155/ha (Blume species); US$2,524/ha (*Avicennia alba* species); US$5,821/ha (*Avicenniaoffcinalis* species) – corresponding to an average value of US$3,072/ha. This corresponds to an annual benefit of US$325/ha. In other words, the economic value of carbon sequestration from mangroves in Ngoc Hien District is about US$325/ha/year. Alternatively, if we use the social cost of carbon value in the range of US$42 per ton of CO_2,[2] the economic value of carbon sequestration from mangroves in Ngoc Hien District is about US$1,720/ha/year.

Recreational benefits (mangrove's cultural services)

The calculation of recreational benefits was estimated by exploring the use of the travel cost method and using the information collected by a questionnaire.

Table 8.9 The value of carbon sequestration for mangrove forests in Ngoc Hien District

Species	Age	Relevant CO_2 content (ton/ha)	Value (US$/ha)	Average value (US$/ha/year)
Rhizopho-	3	40.23	320	107
raapiculata	7	99.96	795	114
Blume species	13	275.66	2,192	169
	16	475.7	3,782	236
	18	463.6	3,686	205
	Average		2,155	166
Avicennia	n/a	171.35	1,362	n/a
alba species	n/a	240.91	1,915	n/a
	n/a	331.29	2,634	n/a
	n/a	526.19	4,183	n/a
	Average		2,524	
Avicenniaoff-	6	537.44	4,273	712
cinalis species	10	926.81	7,368	737
	Average		5,821	724
Average for the mangrove in Ngoc Hien District			3,072	326

Note: n/a – not applicable

According to survey responses, visitors to the Mui Ca Mau resort, Ca Mau National Park, come from 14 different provinces in Vietnam. However, the number of visitors is concentrated mainly in the southwest, including Ca Mau (68%), KienGiang (17%), CanTho (6%). The modes of transport used are automobiles (residents of other provinces) and motorbikes (local people). The average respondent is 38 year old, travels in groups and stays in the region for about 1.3 days.

During this period, the average respondent spends about US$55. The main distinguishing characteristic of Mui Ca Mau Resort from other resorts is that the number of visits by tourists is quite high, at 2.3 visits/year. For other resorts in Vietnam, this number is under 1.5 (Hanley and Barbier 2009). This suggests that this national park, and its well-known mangroves ecosystems, plays a significant role in attracting tourists (Table 8.10). The main purpose for travelling to Ca Mau National Park is to enjoy the natural landscape (88%) and other activities such as scientific research and understanding local cultures (12%). Despite tourists' interest in natural landscapes of the area, 95% of respondents report being unsatisfied with the current infrastructure situation, including the poor conditions of roads, the lack of hotels and other lodging services, while 65% of respondents complained about travel services, information and transportation and guides. Finally, about 35% of respondents expressed their concerns about issues related to the health of the natural environment, in particular those due to the lack of waste treatment systems.

Table 8.10 Some descriptive statistics of the respondents

Variable	Average	Minimum	Maximum
Age	38.9	23	58
Gender (1 = male)	0.6	0	1
Education (study years at school)	15	12	16
Income (US$/month)	157	55	382
Visits (per year)	2.3	1	5
Stay duration (number of days)	1.3	1	3
Group (number of people/group)	16	4	34
Travel cost (US$/visit)	52	12	223

Taking into account the statistical information collected by the questionnaire, the demand in terms of number of visits was estimated. The number of visits was estimated in accordance to the following equation:

$$LnV_i = \beta_0 + \beta_1 age_i + \beta_2 gen_i + \beta_3 educ_i + \beta_4 inc_i + \beta_5 sd_i + \beta_6 gs_i + \beta_7 tc_i + \varepsilon_i$$

With:

age_i age of respondent i
gen_i gender of respondent i
$educ_i$ education of respondent i
inc_i income of respondent i
d_i duration stay of respondent i
gs_i number of people/group of respondent i
tc_i reported travel cost of respondent i (in million Vietnamese Dong – VND)

Results of econometric analysis show that the estimated coefficients have expected signs: young people tend to travel more than older people, men visited study sites more than women or people with high income are able to visit tourist areas more than those with low incomes. However, only length of stay and total travel cost variables are statistically significant. According to valuation literature on travel cost – see Loomis and Walsh (1997), individual consumer surplus from each visited is computed by:

$$CS = -1/\beta_{tc}$$

Taking into account the parameter estimates from the Poisson model, we can infer that the consumer surplus from each visit in accordance to

$$CS = 1/\beta_7 = 1 / 0.0000058 = 172,413$$

In other words, on average a visit to the park generates welfare benefits estimated to be US$7.9, which is the equivalent to 172,413 Vietnamese Dongs. Since a respondent visits the park up to five days, therefore his/her annual welfare benefit is estimated to be up to US$18.2. With an annual number of visits to the park that is estimated to be 60,000 (Department of Statistics of Ngoc Hien District, 2013), the economic value of recreation benefits is estimated to be about US$1.1 million per year. In other words, the economic value of recreation values of the mangroves in Ngoc Hien District is about US$25/ha/year.

Summary and policy impact

Estimation results for the four main types of mangrove ecosystem services in Ngoc Hien District, Ca Mau, are summarized in Table 8.11. Estimation results show that economic value of the four selected ecosystem services of the mangroves in Ngoc Hien District ranges between US$1,390 and 1,560/ha/year. This magnitude is equivalent to 76% of the annual income as reported by individual respondents in the travel cost questionnaire. Furthermore, the great majority of the economic value, 71%, is only captured by the use of other market information, including the social cost of carbon, damages by extreme weather events and travel expenditures incurred with visiting the site. In other words, only 29% of this benefit is currently captured by existing markets. In fact, provisioning services represent 29% of the total economic value of the mangroves.

However, if we take into account the revised social cost of carbon, the economic value of the four selected ecosystem services of the mangroves in Ngoc Hien District is estimated to be in the range of US$2,621–2,985/ha/year. In this scenario, existing market prices only capture 15% of the ecosystem services benefits provided by mangroves. These economic values may be useful in many ways. For example, for policy assessment purposes, the valuation results helps: (1) to determine whether or not a development policy that may alter the status of a mangrove ecosystem creates welfare for community and society, (2) to provide a scientific basis in selecting how to use natural resource in the most efficient way by looking at the trade-offs among the different ecosystem services and (3) to identify the winners and the losers when proposing changes in land use. Furthermore, (4) the economic valuation of the carbon sequestration services shows that the greatest potential of the 43,523 ha mangrove forest area in Ca Mau is in ecosystem-based carbon mitigation, because this area alone is responsible for the offset of 1% of Vietnam's total annual CO_2 emissions.

Table 8.11 Economic values of mangrove's ecosystem services: summary

Mangrove services	Value (US$/ha/year)
Provisioning	352–454
Coastal protection	742–756
Carbon sequestration	325–1,720
Recreational value	25–55

Finally, these results constitute a scientifically based argument for developing future policies on payment for ecosystem services, enabling policy makers to calculate the true value of mangrove ecosystems for a transition to a green economy, as highlighted by the National Green Growth Strategy to 2020 for Vietnam.

Mainstreaming activities

ProEcoServ engaged in wide range of mainstreaming ecosystem services related activities, ranging from capacity-building activities, awareness raising and building partnerships. The target audience and stakeholder groups included, but were not limited to, the following: (1) government ministries, public agencies and policy makers; (2) national and international development partners; (3) scientists including academic institutions and researchers; (4) non-governmental organizations and (5) local communities, civil society and small and medium enterprises. This section reviews these activities.

Capacity building

ProEcoServ engaged in capacity-building activities in Vietnam in a systematic way in order to disseminate long-term impacts from the four-year project. Between 2012 and 2015, 13 workshops and training sessions with more than 600 participants were conducted both at the national and provincial level. This brought together ministries, specialized departments, international organizations, NGOs, and other stakeholders that are focused on raising awareness and improving the technical capacity of target groups on ecosystem services (ESs), the application of mapping tools and policy development. To disseminate the appropriate tools for learning, a number of manuals were produced under ProEcoServ, including a *Manual for Valuation of Ecosystem Services* and a *Manual for Mainstreaming of ESs into Decision Making Processes*. Furthermore, a series of policy briefs have been produced, such as mainstreaming of ESs through the System of Environmental-Economic Accounts, economic incentives for mainstreaming of ESs. Brochures of ProEcoServ were also produced to introduce the project and its results (Figure 8.1).

During May 29–30, 2014, ProEcoServ organized the Policy Dialogue on Mainstreaming Natural Capital into Development Decision: Bringing Environment into Center Stage in collaboration with the Asian Development Bank and the Hanns Seidel Foundation to share approaches of mainstreaming with other partners. Also, the Greater Mekong Sub-region Workshop on Implementation of Sustainable Development Goals – Bringing Natural Capital into Center Stage was organized on May 14–16, 2015 in Ha Noi with participation of more than 100 participants from development partners, government officials, NGOs, research institutes. Such initiatives played a crucial role in expanding the knowledge, tools, and skill set associated with ecosystem services and natural capital more broadly, while targeting the relevant stakeholders so that future initiatives can stem from such strengthened capacities.

Figure 8.1 Illustration of a ProEcoServ brochure.

Communication strategy for awareness raising

ProEcoServ's work in Vietnam was characterized by a communication strategy built on the overall message of how well-managed ecosystems are the cornerstones of ecological infrastructure, and therefore they must be recognised as assets that provide a solid return on investments. See Table 8.12 for information on the multi-pronged approaches deployed in the communication strategy.

Table 8.12 Communication strategy

Communication objectives	Target audiences	Medium
A. Establishing information exchange between scientists and policy makers in integrating ecosystem services into the policy-making process both vertically and horizontally.	Researchers of institutes, policy makers	ProEcoServ database, websites, maps, case studies, project publications and training workshops.
B. Informing about ecosystem services, diffusing tools for integrating ecosystem services into national and local planning processes for policy makers.	Policy makers	ProEcoServpolicy briefs, leaflets, posters, website, policy dialogues, communication strategies.
C. Reinforcing awareness and knowledge about the value of ecosystem services for related target audiences, including governmental agencies and local community.	Policy makers and leaders of enterprises	Multi-media communication, traditional communication, posters, leaflets, maps, case studies, training courses, communication strategies.
D. Enhancing the coordination between related initiatives and stakeholders working on ecosystem services (IPBES, IHDP, GLOBE, TEEB)	Policy makers and social and communicational organizations, researchers and related institutes.	Scientific publication of the project, linkages to relevant websites, case studies and presentations.
F. Enhancing the profile of ProEcoServ: heightening the position, role and image of the project to other organizations; raising awareness of related entities over the world about the project contribution.	The government, intergovernmental organizations, UN agencies, sponsors, policy makers at central and local levels, enterprise community.	Press releases, news, videos, posters, leaflets, online documents and case studies.
G. Building and improving communication capacity from national to local levels	Staffs, scientists, NGOs, social organizations, religious or youth groups, for the purpose of social equality.	Training seminars, presentations, FAQs, sets and lists of communication means.

The main purpose of the strategy was to support the integration of ecosystem services into the planning processes and national policy framework through conveying information and knowledge on ecosystem services and biodiversity management to policy makers and other stakeholders, including the local communities and the private sector. Communication activities included:

- Establishing the project website, connecting project material, content, resources, and allowing easy access for anyone interested in the work (see website: http://proecoserv.com.vn/index.php/en/)
- Publishing communication materials (such as brochures, policy briefs, and case studies) and organizing communication events for different target groups
- Organizing training courses for different target groups at the relevant local to national level
- Organizing policy dialogues and consultation workshops
- Developing training materials and workshop presentations
- Fostering awareness and interest in ecosystems and their role in human welfare through creative mediums such as a photography competition (Box 8.2)

Building partnerships at national and international level

ProEcoServ involved the participation of a number of policy makers at both the national and provincial level. The Project Steering Committee was established with representatives of ministries (i.e. Ministry of Planning and Investment, Ministry of Finance, Ministry of Agriculture and Rural Development and Ministry of Natural Resources and Environment). Moreover, the Steering Committee was chaired by the Deputy Minister of Natural Resources and Environment. At the provincial level, a task force group included different departments (i.e. Department of Agriculture and Rural Development, Department of Planning and Investment, Department of Natural Resources and Environment, Ca Mau National Park) to facilitate implementation of the various aspects of the project. The task force group's members have been invited to consultation workshops and training courses at the provincial level. The project also involved working expert groups at the national and provincial level to participate in conducting project studies and peer review processes. At the provincial level, a peer review group with participation of representation from government was established to contribute to review the land use planning of Ca Mau National Park before submitting to the Provincial People Committee for approval.

Box 8.2 Photo competition

On June 14, 2013, the Ecosystem and Ecosystem Services in Ca Mau photography competition was held by the Project Management Unit (PMU) of Pro-EcoServ in collaboration with the Ca Mau Department of Natural Resources

Figure 8.2 Painting competition.

and Environment (Ca Mau DONRE) and the Ca Mau Association of Art and Literature. Its main aim was to use artistic and creative mediums to raise stakeholders' awareness of the role and importance of ecosystems and ecosystem services to human welfare. The event was able to collect a total of 62 photograph submissions by 23 professional and amateur photographers.

Furthermore, ProEcoServ collaborated with Ca Mau DONRE and the Youth Union of Ngoc Hien District with the participation of more than 100 students. The event contributed to enhance awareness of community in general and children in particular on the importance of mangroves for human well-being.

Source: ProEcoServ-VT

In addition, ProEcoServ worked closely with different initiatives working on natural capital, which brought an opportunity to share project experiences and establish partnerships with similar initiatives working in the same area. The mapping tools developed under the project were replicated under a study done by the World Wide Fund for Nature (WWF) in Ben Tre Province (a Mekong Delta region of Vietnam) on development of a technical guideline on mainstreaming of ecosystem-based adaptations into planning process.

The project team collaborated with the World Bank (WB) in Vietnam to develop the natural capital roadmap for Vietnam, which is a key document in the preparation stage to support Vietnam to become a Core Implementing Country (CIC) of the WAVES global partnership. Furthermore, the project team has worked with the Asian Development Bank under the Core Environment Program to further accelerate activities on natural capital in Vietnam and in the Greater Mekong Sub-region. Lessons learnt from ProEcoServ have been shared with other Greater Mekong Sub-region's countries for wider replication in the region.

Furthermore, ProEcoServ-VT's team also worked with GIZ, the German development agency, to implement a project on mainstreaming of ecosystem-based adaptations into strategic planning to mobilize funds from the German government. The lessons learnt from ProEcoServ on mainstreaming approach can be applied in the project as well. The establishment of partnerships with different initiatives was very important to ensure that the synergy among different initiatives contribute effectively on the sustainability of the project, and ultimately provide an opportunity to scale up project benefits to a wider audience. Finally, ProEcoServ-VT's team participated in different international events to share its experiences on mainstreaming processes in Vietnam at the global level. Some of the important events include the World Forum on Natural Capital, November 20–22, 2013; the Regional Workshop called Valuing and Accounting for the Environment in the Asia Region Workshop, 6th and 7th Annual International Conference of the Ecosystem Services Partnership; 12th meeting of the Conference of the Parties to the Convention on Biological Diversity, October 7–12, 2014 in Pyeongchang, Republic of Korea (the experiences of ProEcoServ were presented at Ecosystem-based Adaptation site event organized by CBD Secretariat and GIZ); 5th Sub-Global Assessment (SGA) Network in Dubai, UAE, October 26–28, 2014; GMS Core Environment Program Technical Workshop-Strengthening Partnerships to Increase Natural Capital Investments in the GMS and WGE 9th Semi-Annual Meeting November, 11–12, 2014 in Bagan, Myanmar and the GMS Environmental Minister meetings on January 27–29, 2015 in Myanmar. All in all, these activities have brought an opportunity for the project team to expand partnerships and develop further collaboration with similar initiatives working on natural capital.

Policy intake

At the provincial level

At the provincial level, the project has support from the Department of Natural Resources and Environment and the Ca Mau National Park to mainstream ecosystem services into land use planning of the Ca Mau National Park through the application of supporting tools (i.e. mapping tool and valuation tools) to assess how development processes impact mangrove ESs. The project organized policy dialogues and consultation workshops to raise awareness for policymakers on the importance of ESs and identify the entry points for mainstreaming of ESs into development processes. Land use planning (LUP) for Ca Mau National Park was

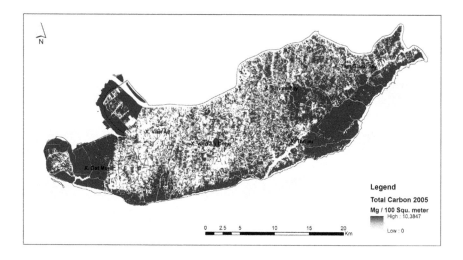

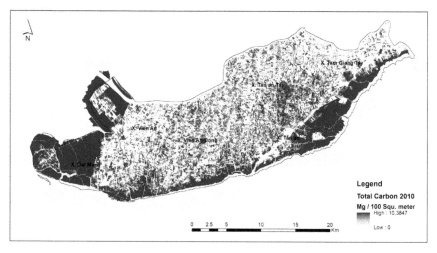

Figure 8.3 Total carbon storage in 2005 and 2010.

identified as an appropriate entry point and received agreement among the relevant agencies. Mapping tools were applied to the mainstreaming of ESs while developing the LUP (Figure 8.3). The Peer Review team, with reviewers from DARD, DONRE, DPI, and the Ca Mau national Park, participated in the review processes of LUP before submitting to the Provincial People Committee for approval.

Moreover, ProEcoServ was among a range of investments focused on mangrove forests including (1) project on restoration of mangrove forests through sustainable shrimp farming and emission reduction in Ca Mau (Project MAM); (2) Program UN-REDD Phase II, which plans to reduce greenhouse gas emissions by 20%from agriculture and rural development sector of Vietnam in 2020; (3) project on mainstreaming adaptation to climate change into coastal areas management plans of Ca Mau carried out by GIZ; and (4) enhancing forest carbon stocks under the mechanism of reducing emissions from deforestation and forest degradation through Project ENRICH, among many other initiatives. It was indicated that interest in acknowledging the value of ecosystems and their services was rising and various international and national entities were coming to collaborate in pushing forward this critical agenda.

At the national policy level

Natural capital in general and ecosystem services in particular have been well recognized under three national legislations of Vietnam, namely, Party Resolution no. 24-NQ/TW, responding to Climate Change, natural resources management and environmental protection, the National Green Growth Strategy to 2020, with a vision to 2050, and the National Strategy for Environmental Protection to 2020, with a vision to 2030. Party Resolution no. 24-NQ/TW was led by the Ministry of Natural Resources and Environment (MONRE) and ISPONRE is assigned as the lead agency to support MONRE in developing the Party Resolution (Box 8.3).

Moreover, at the regional level the project team has participated in contributing to the Greater Mekong Sub-region Ministerial Joint Statement for Promoting Investment on Natural Capital, which was approved at the GMS ministerial meeting on January 2015 in Nay Pyi Taw, Myanmar. The statement has highlighted the importance of strengthening partnerships on natural capital and strengthening the collaboration among GMS countries to enhance investment on natural capital. The commitment to promoting regional collaboration to invest in natural capital has been reflected in the Joint Ministerial Statement. This provided a great foundation to promote regional collaboration in the area, bringing forward key insights of ProEcoServ to the Greater Mekong Sub-region.

Box 8.3 Mainstreaming of natural capital into key national legislation

Guiding principles in party resolution no. 24-NQ/TW

This Party Resolution identified under its guiding principles that "natural resources are the national assets, resources and important natural capital for country development. Therefore, natural resources need to be fully assessed, prized and accounted for in the national economy". The resolution is an important key document to guide different line ministries/sectors for inclusion of natural capital/ecosystem services into their planning processes. Therefore, natural capital – including natural resource and ecosystem services – needs to be fully assessed and accounted for in the national economy. ISPONRE is assigned as the lead agency to support the Ministry MONRE in this mission, and ProEcoServ contributed to this goal with piloting valuation work in Ca Mau.

National Green Growth Strategy in the period 2011–2020

This strategy plans to achieve a low carbon economy and to enrich natural capital and will become the principal direction in sustainable economic development for the country. This strategy brings along with it the implementation of four main activities, which are directly fed by ProEcoServ piloting work. These include: (1) studying and issuing economic and financial policies for restoring and developing natural capital resources; (2) mobilizing and encouraging all economic sectors to invest in ecological services infrastructure, conservation areas and restoration of degraded ecological systems; (3) formulating the green accounting system through valuating natural resources and (4) restoration and development of natural capital.

National Strategy for Environmental Protection to 2020

This strategy plans to achieve the objectives of rehabilitating and regenerating degraded natural ecosystems, especially coastal mangrove ecosystems. This strategy includes working activities such as (1) investigating and conducting assessments to identify the degradation and shrinkage of natural ecosystems in order to develop relevant plans for rehabilitating typical and representative natural ecosystems, especially coastal mangrove ecosystems and (2) increasing the degree of rehabilitation of natural ecosystems to strengthen their resilience to climate change-related impacts, and in the development of relevant mechanisms for payment for ecosystem services towards natural ecosystem rehabilitation, regeneration and conservation. ProEcoServ piloting work in Cau Mau is contributing to these activities.

Source: Party Resolution no. 24-NQ/TW, National Green Growth in the period 2011–2020, National Strategy for Environmental Protection to 2020

Conclusive notes

Inclusion of the ecosystem services narrative into national discourse is crucial for promoting investment in natural capital and providing a legislative framework for sustainable use of natural capital and ecosystem services. The work of ProEcoServ-VT's team played a crucial role in this domain. In addition, ProEcoServ-VT's team contributed in mainstreaming of ecosystem services in policy at the more local, provincial level, with particular attention to the mangrove ecosystem and the Ca Mau National Park. In addition, ProEcoServ contributed significantly to awareness-raising for policy makers on the economic value of ecosystem services and the importance in bringing this information to spatial planning. Furthermore, Pro-EcoServ contributed with the provision of technical capacity to ensure adequate application of the tools for planning processes, including in Cau Mau. Finally, ProEcoServ facilitated the participation of national and provincial government stakeholders in building partnerships, both at national and international scales (including WAVES), and this way contributed to enhance the awareness of the natural capital and ecosystem services in the country as well as the technical capacity in the area of economic valuation of ecosystem services for policy and spatial planning.

Overall, the ProEcoServ-VT has brought positive impacts to mainstream natural capital/ecosystem services to policies at regional, national and provincial levels though mainstreaming of ESs at these levels. Ecosystem services have been successfully mainstreamed into land use planning of Ca Mau National Park which will contribute to conserve the important mangrove ecosystem in Ca Mau.

Notes

1 This is gross income. One should deduct costs of fishing to get net income. However, no systematic information on costs of fishing has been assembled.
2 This value corresponds to the revised Social Cost of CO_2 as indicated by the Interagency Working Group on Social Cost of Carbon, US Government – "Technical Support Document: Technical Update of the Social Cost of Carbon for Regulatory Impact Analysis – Under Executive Order 12866", May 2013. This value is also in agreement with the recent technical discussion of the current values of the Social Cost of CO_2 – see van den Bergh, J.C.J.M. and Botzen, W.J.W. (2014) A lower bound to the social cost of CO2 emissions. *Nature Climate Change* 4, 253–258.

References

Alongi, D.M. (2009) *The Energetics of Mangrove Forests.* Dordrecht, The Netherlands: Springer.

Barbier, E.B. (2007) Valuing ecosystem services as productive inputs. *Economic Policy* 22, 177–229.

Barbier, E.B. and Cox, M. (2004) An economicanalysis of shrimp farm expansion and mangrove conversion in Thailand. *Land Economics* 80(3), 389–407.

Das, S. and Jeffrey, V. (2009) Mangroves protected villages and reduced death toll during Indian super cyclone. *PNAS*, 7359.

de Graaf, G.J. and Xuan, T.T. (1998) Extensive shrimp farming, mangrove clearance and marine fisheries in the southern provinces of Vietnam. *Mangroves and Salt Marshes* 2, 159–166.

Department of Statistics of Ngoc Hien District, 2013.

Donato, D.C., Kauffman, J.B., Murdiyarso, D., Kurnianto, S., Stidham, M., and Kanninen, M. (2011) Mangroves among the most carbon-rich forests in the tropics. *Nature Geoscience* 4, 293–297.

Hanley, N. and Barbier, E.B. (2009) *Pricing Nature: Cost-Benefit Analysis and Environmental Policy*, Edward Elgar, Cheltenham.

Kauffman, J.B., Heider, C., Cole, T.G., Dwire, K.A., and Donato, D.C. (2011) Ecosystem carbon stocks of micronesian mangrove forests. *Wetlands* 31(2), 343–352.

Lee, S.Y. (2004) Relationship between mangrove abundance and tropical prawn production: A re-evaluation. *Marine Biology* 145, 943–949.

Loneragan, N.R., Ahmad, A.N., Connolly R.M., and Manson F.J. (2005) Prawn landings and their relationship with the extent of mangroves and shallow waters in western peninsular Malaysia. *Estuarine, Coastal and Shelf Sciences* 36, 187–200.

Loomis, J.B. and Walsh, R.G. (1997) *Recreation Economic Decisions: Comparing Benefits and Costs*, Second Edition. Venture Publishing, Pennsylvania.

Manson, F.J., Loneragan, N.R., Harch, B.D., Skilleter, G.A., and Williams, L. (2005) A broad-scale analysis of links between coastal fisheries production and mangrove extent: A case-study for northeastern Australia. *Fish Resources* 17, 69–85.

Martosubroto, P. and Naamin, N. (1977) Relationship between tidal forests (mangroves) and commercial shrimp production in Indonesia. *Marine Resources in Indonesia* 18, 81–86.

Murdiyarso, D., Donato, D., Kauffman, J.B., Kurnianto, S., Stidham, M., and Kanninen, M. (2010) Carbon Storage in Mangrove and Peatland Ecosystems: A Preliminary Account from Plots in Indonesia, *CIFOR Working Paper No. 48*, Center for International Forestry Research (CIFOR), Bogor, Indonesia.

Odum, W.E. and Heald, E.J. (1975) Mangrove forests and aquatic productivity. In: Hasler A.D. (ed.) *Coupling of Land and Water Systems, Ecological Studies (Analysis and Synthesis)*, vol 10. Springer, Berlin, Heidelberg.

Pauly, D. and Ingles, J. (1986) The relationship between shrimp yields and intertidal vegetation (mangrove) area: A reassessment. In: Yanez-Arancibia, A. and Pauly, D. (eds.) IOC/FAO Workshop on Recuitment in Tropical Coastal Dermersal Communities, 277–283.

Paw, J.N. and Chua, T.E. (1991) An assessment of the ecological and economic impact of mangrove conversion in Southeast Asia. In: Chou, L.M., Chua, T.E., Khoo, H.W., Lim, P.E., Paw, J.N., Silvestre, G.T., Valencia, M.J., White A.T. and Wong P.K. (eds.) Towards an integrated management of tropical coastal resources, ICLARM Conference Proceedings, 201–212.

Saintilan, N. (2004) Relationships between estuarine geomorphology, wetland extent and fish landings in New South Wales estuaries. *Estuarine Coastal and Shelf Science* 61, 591–601.

Staples, D.J., Vance, D.J., and Heales, D.S. (1985) Habitat requirements of juvenile penaeid prawns and their relationship to offshore fisheries. In: Rothlisberg, P.C., Hill, B.J., and Staples, D.J. (eds.) *Second Australian National Prawn Seminar*, Cleveland, Australia, 47–54.

Turner, R.E. (1977) Intertidal vegetation and commercial yields of penaeid shrimp. *Transaction of the American Fishery Society* 106(5), 411–416.

Yanez-Arancibia, A., Soberon-Chavez, G., and Sanchez-Gil, P. (1985) Ecology of control mechanisms of natural fish production in the coastal zone. In: Yanez-Arancibia, A. (ed.) *Fish Community Ecology in Estuaries and Coastal Lagoons, Towards an Ecosystem Integration*, UNAM Press, Mexico, 571–595.

9 The contribution of forest ecosystems to the Tanzanian economy

Babatunde Abidoye and Eric Mungatana

Introduction and background

One of the key objectives of economic research studies such as this is to generate the factual evidence that policy requires to build a strong business case, which, in the present instance, is for a transformation in forest planning, management and monitoring, in particular to navigate towards a low-carbon development path and a green economy (see for example, UNEP, 2013). NAFORMA (2014: 13) reports that the Tanzanian mainland is estimated to have a total of 48.1 million ha of forests, which is 51% of the total area. Woodlands occupy 44 million hectares, or 91% of the total forest area. NAFORMA categorizes the ownership and management of this forest estate into the following land ownership regimes:

1 Central government land, which is administered by central government agencies such as the TFS or parastatals such as Tanzania National Parks
2 Local government land, which is administered by local government authorities and includes forest reserves decentralized to local government authorities in the 1970s
3 Village land, which is held and administered collectively by village residents under customary law and the Village Land Act, chapter 114
4 Private land, a category that covers all tenure right types giving individual or collective occupancy rights within village, general or government lands (customary right of occupancy, granted right of occupancy, leasehold, and residential licence)
5 General land, which includes land which is not reserved, not occupied or unused forest land
6 Unknown category

Villages are the main owners of forests and woodlands in the Tanzanian mainland, with a 45.7% share, leaving a huge share of the forest estate without official protection status and subject to open access exploitation and heavy pressure (Division of Environment, 2013; see also the forest classification adopted by Ngaga, 2011). The management and development of this vast forest estate is guided by a recently reformulated National Forest Policy, which led to a new Forest Act (2002).

The critical socio-economic importance of forestry to the development aspirations of Tanzania has been extensively demonstrated in current literature. To cite a few illuminating examples, the Statistical Abstract (2012) reports that the 2001 GDP share of forestry and hunting was 3.10% at 1992 prices and 2.70% at 2001 prices (National Bureau of Statistics, Ministry of Finance, 2013). The Tanzanian Ministry of Natural Resources and Tourism (MNRT, 2000) reports that the country's world-famous wildlife and game reserves include 1.6 million ha managed as catchment forests. The Tanzanian Ministry of Natural Resources and Tourism (2000) reports that about 70% of the total forest area of the United Republic of Tanzania is suitable for the production of wood products, with a potential sustained yield of around 16.7 million cubic metres per year (0.7 m^3/ha/yr). The Tanzanian Ministry of Natural Resources and Tourism (2000) demonstrates the substantial provisioning services of *miombo* woodlands, and the forward linkages associated with the primary forest industry. Independently, Ngaga (1998) illuminates the shortcomings of conventional measures of economic performance in capturing the true contribution of forestry to social welfare in the United Republic of Tanzania. Kahyarara, Mbowe and Kimweri (2002) demonstrate the importance of forestry in sustaining rural livelihoods. Kaale (2001) illustrates the huge socio-economic importance of mangrove forests, while the important role of the sector in generating rural and urban employment is demonstrated by the Tanzanian Ministry of Natural Resources and Tourism (Figure 9.1) (1998), Ngaga (1998) and FBD (2000).

Recent studies have, however, brought into sharp public focus the many critical constraints threatening the performance, ecosystem services delivery and sustainable development objectives of this critically vital sector. There exist enormous proximate threats to gazetted forests emanating from a number of sources, including shifting cultivation, wildfires, lack of clearly defined boundaries, illegal logging, expansion of agricultural activities, livestock grazing, unsustainable charcoal production for domestic and industrial use, lack of systematic management, insufficient revenue collection, inadequate infrastructure development, settlement and resettlement, and the introduction of alien and invasive species (see for example Division of Environment, 2009; MNRT, 2000; FBD, 2000) and population pressure. There is evidence of declining capacity of inter-connected industries that depend on the forestry sector for primary inputs. The study by Ngaga et al. (1998) shows that the total installed wood processing capacity of forest-based industries fell from 900,000 m^3/yr of round wood in 1992 to 710,000 m^3/yr in 1998, attributed to obsolete technology, low investment, poor financing and weak market development. In 2000, the Tanzanian Ministry of Natural Resources and Tourism reported that almost all plywood industries in the country were operating below full capacity and generating abundant wastes in production (with some sawmills operating below 35% recovery rate), as a consequence of their very old and poorly maintained machinery. The cumulative effect of all these constraints is the increased deforestation rate currently being witnessed in the country (e.g. see the evidence presented in VPO, 1998; FAO, 2010), threatening future sustainability prospects.

This report is motivated by the hypothesis that the increasing threats faced by the forestry sector could be attributed to the economic characteristics of its outputs.

Figure 9.1 A waterfall in Kilimanjaro. The country's world-famous wildlife and game reserves in Tanzania include 1.6 million ha managed as catchment forests.

Photo credit: Carol Colfer/CIFOR

The sector supplies marketable and non-marketable outputs, with the former being captured within the current system of national accounts (SNA), while a huge proportion of the latter is not captured at all.[1] The report acknowledges that there are credible economic, institutional factors and historical reasons explaining why the sector's non-market benefits typically have provided little incentive for investment and sustainable management. The marketable benefits which are visible and captured in current GDP often present a more compelling case to policy makers. However, restricting the decision maker's attention to the sector's benefits, which are captured by the market and shown only by the GDP share of forestry presents, a skewed picture of their true contribution. As will be shown in this chapter, the well-being of households and the performance of the rest of the economy is intricately linked with the performance of the forestry sector, with the result that losing a country's forest sector goes well beyond losing the sector's GDP share. Consequently, beyond making an attempt to demonstrate the non-market values of forestry, the primary goal of this chapter is to give more visibility to the important role of forestry in supporting the welfare of households and performance of the rest of the economy using data that is currently reported in the SNA.

Guided by these observations and in response to the demonstrated need for sustainable forest management in the United Republic of Tanzania, this study was designed to address three key objectives. First, it used state-of-the-art tools: input-output analysis and social accounting matrix analysis from economics to demonstrate the importance of the sector to the macroeconomy of the United Republic of Tanzania beyond what is reported in the SNA. This analysis will provide the factual evidence required to demonstrate that, in the absence of sustainable forest management, many important welfare-generating upstream and downstream production sectors will fail to perform (i.e. the impacts of failure in the forestry sector goes well beyond losses in the GDP share of forestry) (Figure 9.2).

The output from the first objective then feeds into the second objective, which seeks to answer the question: How do the monetary benefits that society obtains from cutting down its forests (in terms of obtaining useful provisioning forest ecosystem services) compare to the monetary costs of the loss of the value added by forestry to the macroeconomy in the immediate future and long term? This chapter considers the evidence to be generated in response to the first and second objectives as the main contribution (or value added) of this research to forestry in the United Republic of Tanzania.

This report is structured as follows. The section titled "Forest ecosystem services included in the valuation study" looks at the stakeholder engagement process followed to identify the ecosystem services that were included in the study. Since it is hardly feasible to account for all forest ecosystem services within the constraints of a single economic valuation study, stakeholders were chosen based on the importance of their forest ecosystem services that were prioritized for this project. In the section titled "Intersectoral linkages and value added by the forestry and hunting sector", the intersectoral linkages were evaluated using economy-wide models (input-output analysis and social accounting matrices) with a view to demonstrating the critical importance of the forestry sector to the macroeconomy of

Figure 9.2 Lorry with large pieces of wood. Illegal logging a major threat to gazetted forests.
Photo credit: Carol Colfer/CIFOR

the United Republic of Tanzania. Results from this section provide the rationale for investing in sustainable forestry in the United Republic of Tanzania based on data reported in the current SNA. In view of the key conclusions of this section and the potential opportunities provided by internal and external sources of support, the section titled "Towards a sustainable forestry management in the United Republic of Tanzania" uses a stakeholder engagement process again to identify and prioritize investments, policy instruments and institutional arrangements that could in principle be used to support a low carbon development path and a green economy transition in the United Republic of Tanzania. References and appendices appear in the last sections of this report.

Forest ecosystem services included in the valuation study

Ecosystem services

According to the System of Environmental-Economic Accounting 2012 – Experimental Ecosystem Accounting (SEEA Experimental Ecosystem Accounting), ecosystem services are most usefully considered in the context of a chain of flows that connect ecosystems with human well-being (Figure 9.3).

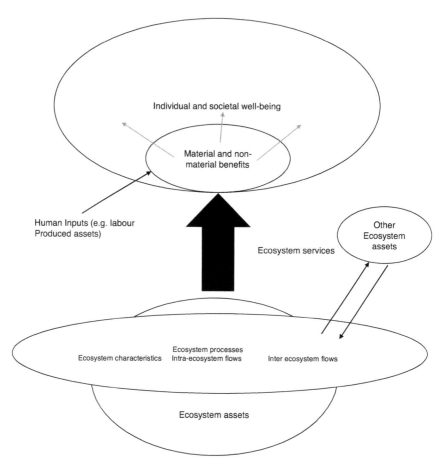

Figure 9.3 Stylized model of flows related to ecosystem services.
Source: Adapted from SEEA Experimental Ecosystem Accounting

Starting with the notion of well-being at the level of both the individual and society, the SEEA Experimental Ecosystem Accounting recognizes that well-being is influenced by the receipt of benefits. In the context of ecosystem accounting, such benefits comprise:

- The products supplied by economic units (e.g., food, water, clothing, shelter, recreation, etc.). These are referred to as "SNA benefits" since the measurement boundary is defined by the production boundary used to measure gross domestic product (GDP) in the System of National Accounts – or SNA.

• The benefits accruing to individuals that are not produced by economic units (e.g., clean air). These benefits are referred to as "non-SNA benefits", reflecting the fact that the receipt of these benefits by individuals is not the result of an economic production process defined within the System of National Accounts.

It follows that, under the SEEA Experimental Ecosystem Accounting, ecosystem services are considered to be "the contributions of ecosystems to benefits used in economic and other human activity", as defined in the SEEA glossary (European Commission et al., 2014), a definition that excludes some flows that are categorized as ecosystem services in other contexts. In particular, these include flows within and between ecosystems that form part of continuing ecosystem processes, commonly referred to as "supporting services" (see the Millennium Ecosystem Services Assessment, 2005). This is the definition and classification of ecosystem services that will be employed here.

Provisioning and intermediate forest ecosystem services selected for inclusion in this study

The ecosystem services identified for inclusion in this study were selected on the basis of two stakeholder consultations. The first was held on June 4, 2013, between the Centre for Environmental Economics and Policy in Africa and the Tanzania Forest Services, the institution formally tasked with coordinating this research project given its role as chief custodian of forests in the United Republic of Tanzania. In the meeting the Tanzania Forest Services highlighted the important connection that exists between effective forest management and the performance of the following sectors in the United Republic of Tanzania: rural and urban households, forestry, domestic and industrial water supply, rain-fed and irrigated agriculture, livestock and inland fisheries, domestic energy supply, hydroelectric power generation, tourism, wildlife and beekeeping.

  Consensus emerged from the meeting that the current study should generate an improved understanding of the role of better-managed forests in the macro-economy of the United Republic of Tanzania by considering the listed ecosystem services. It also emerged from the meeting that the Tanzanian Ministry of Natural Resources and Tourism, through the ministry's Forestry and Beekeeping Division and with assistance from, among others, the Norwegian Agency for Development Cooperation (NORAD), had prepared a major research study in 2003 entitled "Resource Economic Analysis of Catchment Forest Reserves in Tanzania" (hereinafter referred to as the 2003 MNRT study). This study estimated monetary values for the following forest ecosystem goods and services of relevance to the current study: timber and timber-related values, non-timber forest products, water (domestic and livestock use, irrigation, electricity and fisheries), measures for erosion protection, tourism, carbon sequestration, biodiversity, option values and non-use values.

The second consultation took place at a stakeholders' workshop in Dar es Salaam on July 17, 2013, and was attended by representatives of six government ministries (Natural Resources and Tourism, Agriculture, Energy and Mining, Water, Livestock Development, and Fisheries and Works), two public institutions (National Bureau of Statistics and the National Environmental Management Council), the Sokoine University of Agriculture, and four international organizations – the United Nations Environment Programme (UNEP), the United Nations Development Programme (UNDP), the International Centre for Research in Agroforestry (ICRAF), and the Food and Agriculture Organization of the United Nations (FAO). There was consensus at the second stakeholders' meeting of the importance of a study that provided a better understanding of the importance of forests to the macro-economy of the United Republic of Tanzania. Stakeholders at that meeting also concluded that any forest valuation work to be delivered under the current study should add value to what is currently known from the 2003 MNRT study.

Intersectoral linkages and value added by the forestry and hunting sector

The forestry sector provides both use and non-use values for the economy, as indicated in the previous section. The use-values can be estimated using market mechanisms and can be observed through the activities of households, private enterprises and the government. In the present section, we evaluate the importance of the forestry sector by looking at its linkage with other sectors of the economy and investigate the value which it adds using input-output analysis and social accounting matrices. The input-output model is based on an analysis of inter-industry transactions and examines how industries use the products of other industries as inputs for their own products. One of the main advantages of the input-output model is that its analysis of inter-industry transactions can be used to estimate the economic impacts of any changes to the economy.

The social accounting matrices are an extension of the input-output tables. In addition to the income and expenditure flows of industries and their outputs, as captured in the input-output tables, the social accounting matrices contain detailed information on different institutions. The matrices thus incorporate institutional and structural details that capture all transfers and real transactions between industries and institutions in the economy. Since the social accounting matrices incorporate the input-output table, they provide a comprehensive economy-wide database with an internally consistent set of accounts for production, income and expenditures.

Data were available for the input-output table of the United Republic of Tanzania for the years 2000–2010 and these were used in the computation of the social accounting matrices, which then served as the major data source for analysing inter-industry linkages. The following framework (see Box 9.1) is based on that prepared by Parra and Wodon (2009).

Contribution of forestry and hunting, and the wood,
paper and printing sectors to value addition

Box 9.1 Key messages for analysts and policy makers

Using the data on intersectoral transactions captured in the social accounting matrices for the United Republic of Tanzania, the section titled "Intersectoral linkages and value added by the forestry and hunting sector"
demonstrates the following:

- Each hectare of forest that is left standing boosts the contribution to
  the GDP of the forestry and hunting sector. Thus, in 2001, forestry
  and hunting contributed TSh 296.7 billion to the GDP in 2001of
  the United Republic of Tanzania, which translates into TSh 6,168
  per ha per year (2001) equivalent to TSh 29,233.84 per ha per year
  (2013).
- For each hectare of forest that is cut down, there are two consequences:
  first, the cleared hectare will no longer contribute to the current GDP,
  which is estimated at TSh 6,168 per ha per year (2001); and, second,
  the potential value added by forestry to other sectors of the Tanzanian
  economy – in terms of income and valued added taxes – to a total
  quantity estimated in the present study at TSh 10,599 per ha per year
  (2001) will be lost. Accordingly, clearing a hectare of forest translates
  into a total loss of TSh 16,767 per ha per year (2001) in terms of direct
  losses and losses incurred by other sectors, which is equivalent to TSh
  83,771.70 per ha per year (2013).
- An increase in the consumption of the forestry sector by households
  resulted in an increase in GDP, household income, wage rates and composite commodity prices.

These predictions are clearly of importance in informing forest policy:
in the interests of improving the welfare of rural poor, rural non-poor,
urban poor and urban non-poor households, forest policies should
encourage growth in sectors that make use of forestry as an input in
their production.

"Value added" is defined as the sum of factor incomes and value added taxes.
In this section we consider the question: What is the contribution of the 17 industries modelled in the social accounting matrices to aggregate GDP? Table 9.1
presents the contribution to GDP by sector (in descending order), highlighting
the value added by the sectors of forestry and hunting, and that of wood, paper
and printing.

Table 9.1 Value added (in billions of Tanzanian shillings and US dollars)

Activity	Value added in TSh billion	Value added in US$billion*
Agriculture (AAGRIC)	3,224.0	3.68
Real estate (AESTAT)	1,879.3	2.14
Food processing (APFOOD)	1,347.7	1.54
Public administration (AADMIN)	653.3	0.75
Livestock and fishery (ALIVES)	622.2	0.71
Construction (ACONST)	582.0	0.66
Textiles (ACLOTH)	305.8	0.35
Forestry and hunting (AFOREST)	296.7	0.34
Other manufacturing (AOTHM)	263.4	0.30
Hotels and restaurants (AHOTEL)	259.6	0.30
Trade (ATRADE)	253.6	0.29
Transport and communications (ATRANS)	238.9	0.27
Private services (APRIVS)	224.8	0.26
Utilities (AUTILI)	112.0	0.13
Machinery and equipment (AEQUIP)	75.7	0.09
Wood, paper and printing (AWOODP)	72.0	0.08
Mining (AMININ)	21.1	0.02
Aggregate	**TSh** 10,432 billion	**US$** 11.90 billion

Source: Author's computation

*The mean exchange rate in 2001 was 876.71 Tanzanian shillings to the dollar

Agriculture (AAGRIC) is the major contributor (TSh 3,224 billion), while forestry and hunting (AFOREST), and wood, paper and printing (WOODP) contribute TSh 296 billion (eighth overall) and TSh 72 billion (16th overall), respectively. These sectors, which are the two sectors of central interest to this study, make up 3.5% of the GDP of the United Republic of Tanzania. This result is based on current measures of economic performance; other sectors contribute more to the national income by several orders of magnitude (for example, the contribution of agriculture is more than ten times that of forestry and hunting).

The remaining parts of this section will explore the critical role played by the forestry and hunting sector in the macro-economy of the United Republic of Tanzania by carrying out the following analyses: multiplier (in the section titled "Multiplier analysis, sectoral growth and price impacts"), forward and backward linkages (in the section titled "Forward and backward linkages"), structural path analysis (in the section titled "Structural path analysis"), simulating data to capture household demand for forestry output that is not presently captured in the system of national accounts (in the section titled "Simulating data to capture higher household demand for forestry output"), and finally, the cost-benefit of deforestation based on the representation of forestry in the system of national accounts of the United Republic of Tanzania (in the section titled "Cost-benefit analysis of deforestation based on the representation of the forestry sector in the national accounts of the United Republic of Tanzania").[2]

Multiplier analysis, sectoral growth and price impacts

We begin here by assuming that the Tanzanian Planning Commission proposes a policy that would result in increased activity (or an increased supply of goods and services) from the forestry and hunting (AFOREST), agriculture (AAGRIC), and wood, paper and (AWOODP) sectors by the same proportional amount (e.g., a targeted 10% supply increase in each sector within the next five years). Many factors could potentially drive such an increase in supply, including increased exogenous demand,[3] and options for making each of these policies operational could include increasing the annual budgetary allocation for each sector.

In this section, multiplier analysis[4] is employed to assess whether household incomes would increase or decrease as a result of these proposals; thus, it considers the question: Would households benefit from or be hindered by the individual proposals?

As noted previously in this section: the present section distinguishes between four types of households: first, rural poor (RURPOOR); second, rural non-poor (RURNPOOR); third, urban poor (URBPOOR) and, fourth, urban non-poor (URBNPOOR). It follows that, by answering the previous question, the analysis will make it easier to predict how the proposed policies would affect welfare distribution across households. In the present section, the multipliers are presented in absolute values and in what might be termed "elasticity values".[5] The section will further deconstruct the multipliers into transfer, open-loop and closed-loop effects to facilitate a better understanding of their policy significance.[6] Transfer effects are designed to capture the impact of the proposed policies, based on transfers within the group of accounts. Open and closed-loop effects are also called direct and indirect effects, respectively. The direct effect of the forestry and hunting sector, for example, focuses on the sector's impact on final demand, i.e., the goods and services supplied by the sector and directly consumed as final products, such as the direct gathering of firewood by households for domestic energy supply. The indirect effect of the forestry and hunting sector includes output that helps support the production activities of other sectors in the economy. The forestry and hunting sector, for example, indirectly contributes to the value added in the electricity generation sector.

Impacts of exogenous increases on households

Using the constructed model, an analyst can sequentially assess the impact of a unit increase in each of the activities or sectors on the rest of the economy. The full analysis for all activities is presented in Annex I of the full report (UNEP, 2014). Table 9.2 abstracts data from this Annex to highlight the key results which are relevant to forest policy analysis.

Based on the multiplier analysis (Table 9.2), it may be predicted that a one-unit exogenous increase in the demand for forestry and hunting will lead to increases of 0.5 units in the income for the rural poor, 1.9 in income for the rural non-poor, 0.1 in the income for the urban poor and 0.7 in the income for the urban non-poor. By comparison, the impacts of a similar increase on the exogenous demand for output in agriculture on the one hand, and the wood, paper and printing industries on the other, are given in columns 3 and 4. To give a more practical interpretation to the multipliers reported in Table 9.2 let us assume an annual income of the rural poor in the United Republic of Tanzania of TSh 250,000, of the rural non-poor of TSh 600,000, of the urban poor of TSh 300,000 and of the urban non-poor of TSh 900,000. We can use the multipliers of Table 9.2 to derive the predictions of Table 9.3.

Thus, if the Planning Commission were to increase output from the forestry and hunting sector by a small percentage within the next five years as postulated, the model predicts that the annual income of the rural poor would increase from TSh 250,000 to TSh 380,825, of the rural non-poor from TSh 600,000 to TSh

Table 9.2 Predicted impacts of exogenous increases on changes in household welfare (multipliers)

Impact of exogenous demand on incomes of:	*Exogenous increase in demand for output from:*		
	Forestry and hunting*	*Agriculture**	*Wood, paper and printing**
Rural poor households	0.52	0.50	0.33
Rural non-poor households	1.87	1.79	1.50
Urban poor households	0.09	0.09	0.11
Urban non-poor households	0.71	0.72	0.78

Source: Author's computation

*Values have been rounded off

Table 9.3 Predicted impacts of exogenous increases on household welfare (Tanzanian shillings)

Household sector	*Hypothesized current annual income (TSh)*	*Estimated total income (TSh) after an exogenous increase in the demand for output from:*		
		Forestry and hunting	*Agriculture*	*Wood, paper and printing*
Rural poor	250,000	380,825	373,775	333,675
Rural non-poor	600,000	1,721,460	1,074,060	1,498,200
Urban poor	300,000	326,040	326,640	334,170
Urban non-poor	900,000	1,540,890	1,548,990	1,597,770

Source: Author's computation

1,721,640, of the urban poor from TSh 300,000 to TSh 326,040 and of the urban non-poor from TSh 900,000 to TSh 1,540,890. The analyst can use the predictions from Table 9.3 to address two questions of policy interest. First, which of the three sectors of investment would bring the greatest benefit to the poor? Second, what would be the likely impact of the chosen investment on the welfare of the rural poor, rural non-poor, urban poor and urban non-poor before and after the policy is implemented? According to the predictions of Table 9.3:

- Increasing the exogenous demand for output from the forestry and hunting sector has a larger impact on the incomes of the rural poor and non-poor compared to similar increases in the demand for output from the agriculture and wood, paper and printing sectors. It follows that adopting such a policy would benefit the rural poor and non-poor much more than adopting similar policies in the agriculture or wood, paper and printing sectors.
- The model predicts that such a policy would enhance the welfare of the rural poor and non-poor. The Ministry of Natural Resources and Tourism could potentially use these predictions to argue for additional funding support to the sector in the interests of poverty alleviation. The ministry could also use this prediction to seek pro-poor donor funding for rural development and forest conservation.
- In all cases, the model predicts that the rural non-poor consistently gain much more than the rural poor. The government should thus consider supporting complementary investments designed to reduce incidences of overall rural poverty, such as the provision of safer drinking water to save time and effort spent collecting water, the improvement of sanitation to reduce vulnerability to diseases, educational programmes to facilitate more efficient use of resources through environmental awareness, etc. It is clear from the predictions that investments in forestry should be viewed as a single ingredient in an overall rural development strategy.

Lastly, it may be seen from the data in Annex I of the full report (UNEP, 2014) that:

- Of all the production sectors in the United Republic of Tanzania captured in the analysis, an exogenous increase in demand for forestry and hunting has had the largest impact on household incomes (rural poor and non-poor, urban poor and non-poor).
- If the government target is growth in rural incomes, implementing policies that increase the output from the forestry and hunting sector appears to be most promising (of all sectors).

Deconstructing the multiplier

The preceding analysis derived quantitative estimates of the total impact (referred to as multipliers) of a unit increase in the sectors of forestry and hunting, agriculture, and wood, paper and printing on the welfare of rural poor, rural non-poor, urban

poor and urban non-poor households. The objective of this section is to deconstruct these multipliers into three kinds of effects: transfer, open-loop (or direct) and
closed-loop (or indirect). It was stated earlier that the transfer effect is designed to
capture the impact of the exogenous increases based on transfers within the group
of accounts. Since the results (see Annex II of the full report, [UNEP, 2014]), show
very minimal transfer effects on most activities in the United Republic of Tanzania,
the rest of the analysis will concentrate on the open- and closed-loop effects. A
logical question would then be: What is the value added by deconstructing the multipliers into open and closed-loop effects? Value is added because an understanding
of the relative strengths of the two sources of impact helps answer the following
policy-relevant question: Which particular activities – both direct and indirect – will
help uplift the welfare of the rural poor, and should forest policy encourage investments that enhance the operation of the direct effect or the indirect effect?

The full results for this section are available in the full report (UNEP, 2014). The
summary provided in Table 9.4 shows that a one-unit exogenous increase in the
demand for output from forestry and hunting directly benefits the rural non-poor
disproportionately more than the other household sectors (rural poor, urban poor
and urban non-poor).

This result reinforces our earlier conclusion that the rural non-poor are more
likely to benefit from increased investments in the forestry and hunting sector than
the rural poor (see Tables 9.2 and 9.3). A similar pattern emerges with an exogenous increase in the demand for output from the agriculture (column 3) and the
wood, paper and printing sectors (column 4). Table 9.4 further shows that exogenous increases in the demand for output from forestry and hunting have much
larger direct impacts on the welfare of rural areas (rural poor and rural non-poor)
compared to similar increases in the demand for output from the agriculture or the
wood-paper-printing sectors.

The value of these predictions in informing forest policy could be summarized
as follows: in the interests of improving the welfare of rural poor, rural non-poor,
urban poor and urban non-poor households, forest policy should encourage growth
in sectors that make use of forestry as an input in their production.

Table 9.4 Open-loop (direct) effect (multipliers)

Impacts on incomes of:	Exogenous increase in the demand for output from:		
	Forestry and hunting*	*Agriculture**	*Wood, paper and printing**
Rural poor	0.29	0.18	0.06
Rural non-poor	0.61	0.57	0.44
Urban poor	0.02	0.02	0.05
Urban non-poor	0.16	0.19	0.31

Source: Author's computation

*Values have been rounded off

Forward and backward linkages

This section focuses specifically on the economic relationships that exist between the forestry and hunting sector and other productive sectors in the economy, by studying forward and backward linkages. Forward linkages are said to exist when the growth of one sector leads to the growth of other sectors that use its output as their input.[7]

Backward linkages, by contrast, are said to exist when the growth of one sector leads to the growth of other sectors that supply its inputs. The figure classifies activities according to the size of their forward and backward linkages.[8]

For an even better picture of the economic interlinkages, we present the economic landscape of the United Republic of Tanzania in Figure 9.4. Economic landscapes enable us to visualize, in a simple picture, complex relations in the economy, and also those between individual sectors and the economy as a whole. The axes of the economic landscapes are the sectors or agents involved in the productive processes, while the heights are the values resulting from the transactions and interactions, either directly or indirectly. The heights could include, for example, values of production, value added, imports and number of people employed. In presenting the economic landscape of the United Republic of Tanzania, the report identifies sectors based on the first order change in the sum of all cells of the inverse matrix caused by changes in the technical coefficients.[9]

Our results show that the forestry and hunting sector (AFOREST, activity 3) ranks fourth of all the country's industries, based on the backward linkage hierarchy only behind those of food processing (AFOOD, activity 5), public administration (AADMIN, activity 16) and real estate (AESTAT, activity 15).

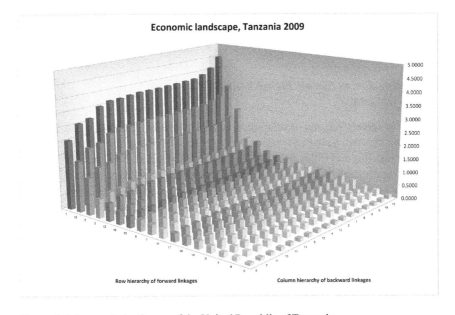

Figure 9.4 Economic landscape of the United Republic of Tanzania.

Structural path analysis

It is important for policy makers to know the path through which the value of increased activity in a sector passes to households and other agents in the economy. This helps them see in detail how the effect of a change in a sector unfolds before getting to the final household type. The objective of this section is to apply structural path analysis to the social accounting matrices framework, to identify the path through which influence of a particular sector is transmitted.[10] In the section titled "Multiplier analysis, sectoral growth and price impacts", we highlighted the fact that a one-unit exogenous increase in the demand for forestry and hunting leads to increases of 0.5 units in the income of the rural poor; 1.9 units in the income of the rural non-poor; 0.1 units in the income of the urban poor and 0.7 units in the income of the urban non-poor.

This impact, however, has a channel through which it passes before reaching these agents. Structural analysis seeks to evaluate which other agents are affected before the final increase of 0.5 units for the rural poor. That is, while we know that a shock in the forestry sector has a final impact on the rural poor, we need to ascertain which other sectors or accounts were affected before the final impact on the rural poor. Thus, we consider how an exogenous increase in forestry and hunting (AFOR-EST) affects different household agents. In other words, we endeavour to answer the question: Does the increase of 0.5 units for the rural poor affect only this group or does it first have an impact on the urban rich before trickling down to the rural poor?

Starting with the influence of forestry on the rural poor, we find that 54.5% of the multiplier travels through the path connecting forestry to the rural poor directly. At the same time, 5.1% of the multiplier travels through the rural non-poor, onward to subsistence labour, then to agriculture, before influencing the rural poor. This makes sense, given the level of communal living and dependence in many African countries. In the case of the rural non-poor, the majority (74.5%) of the multiplier travels directly to the rural non-poor, while a mere 2.1% travels through livestock consumption to the rural non-poor.

Simulating data to capture higher household demand for forestry output

The use of forests in many developing countries is usually undervalued by the relevant ministries and managed sustainably in a pluralist and intersectoral manner (see, for example, Roe and Elliot 2010 and Hassan and Mungatana 2013, section 5). In the United Republic of Tanzania for example, forests are a source of income for some households or consumed as a supplement to other goods. For instance, in many communities, wood is the source of fuel for cooking food. Data on these activities, however, are typically not available at the national level. As a result, the contribution of the forestry sector to the economy is undervalued, an issue that was briefly highlighted in the section titled "Impacts of exogenous increases on households".

The study by Agrawal et al. (2012) observes that in many developing countries the non-industrial economic contributions of forests are typically unrecorded and in many cases are between three and ten times higher than the revenues collected in national accounts. One way of dealing with this issue and analysing the situation as

Table 9.5 Percentage change in household income

Household type	Total income*	Capital income*	Labour income*
Rural (below food poverty line)	1.09	0.94	1.22
Rural (between food and basic needs poverty lines)	1.07	0.94	1.24
Rural (non-poor – head with no education)	1.16	0.94	1.39
Rural (non-poor – head not finished primary school)	1.03	0.94	1.14
Rural (non-poor – head not finished secondary school)	0.58	0.94	0.94
Rural (non-poor – head finished secondary school)	0.21	0.94	0.22
Urban (below food poverty line)	0.30	0.94	0.67
Urban (between food and basic needs poverty lines)	0.17	0.94	0.63
Urban (non-poor – head with no education)	0.30	0.94	0.99
Urban (non-poor – head not finished primary school)	0.39	0.94	0.74
Urban (non-poor – head not finished secondary school)	0.29	0.94	0.57
Urban (non-poor – head finished secondary school)	0.32	0.94	0.23

*Values have been rounded off

Source: Author's computation

it would have appeared if the data had been adequately captured is to use a comput-able general equilibrium (CGE) model. Consequently, in this section we use a CGE analysis to model the contribution of the forestry and hunting sector, by simulating an increase in the use of the forestry sector greater than that currently reported in the national accounts. We highlight in the following our key findings, with the full analysis presented in Annex XII of the full report (UNEP, 2014). If household consumption of the forestry sector is factored into the calculations, in order to capture the contribution of this sector to household demand, the resulting figures show a 0.6 percentage point increase in GDP at market prices, a 0.8 percentage point increase in the consumer price index, and a 0.6 percentage point increase in the GDP deflator. As expected, total income increased for all household types that experienced the shock (Table 9.5). The impact is highest, however, on non-poor rural households with a non-educated household head (a 1.16 percentage point increase in total income). This is followed by households that are in the rural areas below the food poverty line. Capital income increases by about the same rate for all the households but labour income varies by household type. In practical terms, this exercise tells us that if household consumption of the forestry sector were to be appropriately recorded and captured, the sector would show a higher contribution to GDP at market prices and demonstrate that it contributes more to the income of households in the rural area and to the less educated urban non-poor.

Wages increases can be seen in all industries that are based on this scenario (Table 9.6). With increased consumption by households of forestry commodities, labour becomes significantly more expensive in the meat-processing and dairy-products indus-try (1.53 percentage point increase). This is as a result of the linkage between this sector and that of forestry and hunting. The wage rate in the forestry and hunting sector also

Table 9.6 Summary of results of a 10% shock on the forestry and hunting sector

Activity or sector	GDP before shock (TSh billion)	GDP after shock (TSh billion)	Increase in GDP after shock (TSh billion)	Increase in GDP after shock (percentage)
Agriculture (AAGRIC)	3,223.96	3,235.93	11.97	0.37
Forestry and hunting (AFOREST)	**300.53**	**311.43**	**10.9**	**3.63**
Real estate (AESTAT)	2,032.60	2,040.11	7.51	0.37
Food processing (AFOOD)	1,570.49	1,576.64	6.15	0.39
Livestock (ALIVES)	629.72	632.58	2.86	0.45
Trade (ATRADE)	1,013.36	1,015.74	2.38	0.24
Textiles (ACLOTH)	412.33	413.94	1.61	0.39
Hotels and restaurants (AHOTEL)	453.81	455.16	1.35	0.30
Other manufacturing (AOTHM)	381.80	383.05	1.25	0.33
Public administration (AADMIN)	1,585.08	1,586.17	1.09	0.07
Private services (APRIVS)	401.92	402.87	0.95	0.24
Transport and communications (ATRANS)	684.59	685.48	0.89	0.13
Construction (ACONST)	769.60	770.43	0.83	0.11
Utilities (AUTILI)	216.43	217.15	0.72	0.34
Mining (AMININ)	128.20	128.46	0.26	0.20
Wood-paper printing (AWOODP)	**144.86**	**145.03**	**0.17**	**0.12**
Machinery and equipment (AEQUIP)	115.04	115.13	0.09	0.07
Total	**14,064.33**	**14,115.31**	**50.98**	**0.36**

Source: Author's computation

becomes higher with an increase of 1.27 percentage points. There is no significant impact on the price of labour in the utilities and manufacturing sectors, however.

Cost-benefit analysis of deforestation based on the representation of the forestry sector in the national accounts of the United Republic of Tanzania

For the purposes, however, of providing information and making policy choices that have an impact on the forestry sector, a much more pertinent question would be: What are the economy-wide magnitudes of the costs and benefits if the current rate of deforestation continues for the next 20 years? Are the net benefits of such a small magnitude such as to justify their being ignored by development policy? Or are they of such a colossal amount that they necessitate the immediate attention of policy makers?

To answer this question, a cost-benefit analysis was conducted of deforestation, using a discount rate of 5%, the rate used by the Bank of Tanzania in analysing long-term investments (Table 9.7).

Table 9.7 Cost and benefit analysis of deforestation based on the representation of forestry in the Tanzanian SNA

Year	Time	Discount factor	Area deforested annually (ha)	Undiscounted			Discounted		
				Benefits (million TSh)	Costs (million TSh)	Net Benefits (million TSh)	Benefits (million TSh)	Costs (million TSh)	Net Benefits (million TSh)
2013	0	1.0000	372,816	10,898.84	31,231.43	−20,332.59	10,898.84	31,231.43	−20,332.59
2014	1	0.9524	372,816	10,898.84	31,231.43	−20,332.59	10,379.85	29,744.22	−19,364.37
2015	2	0.9070	372,816	10,898.84	31,231.43	−20,332.59	9,885.57	28,327.83	−18,442.26
2016	3	0.8638	372,816	10,898.84	31,231.43	−20,332.59	9,414.83	26,978.88	−17,564.05
2017	4	0.8227	372,816	10,898.84	31,231.43	−20,332.59	8,966.50	25,694.17	−16,727.67
2018	5	0.7835	372,816	10,898.84	31,231.43	−20,332.59	8,539.53	24,70.64	−15,931.11
2019	6	0.7462	372,816	10,898.84	31,231.43	−20,332.59	8,132.88	23,305.37	−15,172.49
2020	7	0.7107	372,816	10,898.84	31,231.43	−20,332.59	7,745.60	22,195.59	−14,449.99
2021	8	0.6768	372,816	10,898.84	31,231.43	−20,332.59	7,376.77	21,138.66	−13,761.90
2022	9	0.6446	372,816	10,898.84	31,231.43	−20,332.59	7,025.49	20,132.06	−13,106.57
2023	10	0.6139	372,816	10,898.84	31,231.43	−20,332.59	6,690.94	19,173.39	−12,482.45
2024	11	0.5847	372,816	10,898.84	31,231.43	−20,332.59	6,372.33	18,260.37	−11,888.04
2025	12	0.5568	372,816	10,898.84	31,231.43	−20,332.59	6,068.88	17,390.83	−11,321.95
2026	13	0.5303	372,816	10,898.84	31,231.43	−20,332.59	5,779.89	16,562.69	−10,782.81
2027	14	0.5051	372,816	10,898.84	31,231.43	−20,332.59	5,504.66	15,773.99	−10,269.34
2028	15	0.4810	372,816	10,898.84	31,231.43	−20,332.59	5,242.53	15,022.85	−9,780.32
2029	16	0.4581	372,816	10,898.84	31,231.43	−20,332.59	4,992.89	14,307.48	−9,314.59
2030	17	0.4363	372,816	10,898.84	31,231.43	−20,332.59	4,755.13	13,626.17	−8,871.04
2031	18	0.4155	372,816	10,898.84	31,231.43	−20,332.59	4,528.69	12,977.30	−8,448.61
2032	19	0.3957	372,816	10,898.84	31,231.43	−20,332.59	4,313.04	12,359.34	−8,046.30
2033	20	0.3769	372,816	10,898.84	31,231.43	−20,332.59	4,107.66	11,770.80	−7,663.14

Source: Author's computation

The analysis shows that the present value of benefits to the Tanzanian economy from deforestation amounts to TSh 147 billion (US$92 million) for the period 2013–2033.[11] The present value of costs from deforestation to the Tanzanian economy amounts to TSh 420 billion (US$263 million). The present value of net losses from deforestation to the Tanzanian economy from this scenario therefore amounts to TSh 273 billion (US$171 million) for the period 2013–2033. These are potential real (as opposed to hypothetical) losses to be experienced by production sectors that have economic linkages with the forestry sector which include both public and private production units. Such losses will have potential implications for their net profits.

The question is whether net losses of TSh 273 billion for the period 2013–2033 are of a magnitude that warrants the attention of these production units. Stated otherwise, are the potential net losses of TSh 273 billion colossal enough to spur the private and public sector into action to protect forests in the United Republic of Tanzania? Another way to ask this question would be: suppose the United Republic of Tanzania had TSh 273 billion (US$171 million) today for investment. What would be the economic impact of this investment for the next 20 years? It should be noted that the potential net losses of TSh 273 billion are likely to represent a lower limit, considering that, with population growth among other drivers, the rate of deforestation is likely to increase (and not remain constant as we have assumed here).

Towards a sustainable forestry management in the United Republic of Tanzania

As shown in the section titled "Intersectoral linkages and value added by the forestry and hunting sector", the forestry and hunting sector has emerged as that with the greatest predicted impact on increasing the levels of household incomes (rural poor and non-poor, and urban poor and non-poor). Investing in and expanding this sector could be an important component of a poverty alleviation strategy. It was further predicted that the GDP of all production sectors responded positively to increased activity in the forestry and hunting sector, leading to the conclusion that keeping this sector healthy and vibrant is in the business interests of all production sectors sharing an economic relationship with it, including those of agriculture, real estate, food processing, livestock, trade, textiles, hotels and restaurants, other manufacturing sectors and public administration.

It was further shown that the levels of monetary losses attributed to the current rate of deforestation are large enough to compel the Tanzania Forest Services to take action to avert deforestation, not only in the interests of business but also in view of the socioeconomic role that this sector plays in the country's economy. For example, it was shown that the revenue lost to the Tanzania Forest Services through deforestation would be sufficient for it to fund the annual budget of the West Kilimanjaro plantation, or equivalent to the combined annual budgets of the plantations of Rubya, Kiwira and Kawatire.

In light of these results, a second stakeholder workshop was organized to identify: first, policy options for sustainable forest management in the United Republic of Tanzania; second, potential roles of the household sector in supporting sustainable forestry in the United Republic of Tanzania; third, potential roles of the private sector in supporting sustainable forestry in the United Republic of Tanzania;

and, fourth, the potential role of REDD+ investments in the transition to a green economy in the United Republic of Tanzania.

Policy options for sustainable forest management in the United Republic of Tanzania

The workshop that included policy makers observed that a critical constraint on sustainable forest management was the lack of effective implementation and enforcement of existing forest policy and laws, a situation that must be addressed forthwith. It was earlier observed that the agricultural expansion and demand for rural energy fuelled by population growth are major proximate driving forces behind deforestation. Accordingly, in the interests of sustainable forest management there is a need to integrate forestry with agricultural development, within multiple land-use models, and to provide affordable alternatives to firewood and charcoal as sources of household energy, in particular in the rural areas. Other aspects considered important in a strategy for sustainable forest management include:[12]

- Investment in an effective forest extension education programme
- Investment in improved alternative livelihood for communities that depend on forest resources
- Investment in the establishment of plantations to meet current and future demand for forest products
- Involvement of local communities in forest management and conservation

Figure 9.5 Raking over coals: affordable alternatives to charcoal need to be provided as sources of household energy.

Photo credit: Ollivier Girard CIFOR

Potential roles of the household sector in supporting sustainable forestry in the United Republic of Tanzania

Given the demonstrated importance of the forestry and hunting sector to household welfare, the following strategies were suggested to motivate increased household involvement in sustainable forestry:

- Providing incentives for households to plant trees, especially on farms and in villages, through the adoption of conservation agriculture and other agro-forestry practices
- Involving households in the protection and management of forests under the control of the Tanzania Forest Services by encouraging their participation in the implementation of forest policy
- Assisting households to use improved and alternative domestic energy sources to reduce the impact of fuelwood extraction and charcoal burning

Potential roles of the private sector in supporting sustainable forestry in the United Republic of Tanzania

Given the demonstrated importance of the forestry and hunting sector to supporting economy-wide production, some of the following strategies were suggested to motivate increased private sector involvement in sustainable forestry:

- Providing necessary facilities and incentives to support the establishment of plantations, afforestation, and tree-planting programmes at household and national levels
- Supporting investments in campaigns to raise public awareness of the values of forests and what households and production sectors stand to lose if forests disappear

Potential roles of REDD+ investments in a green economy transformation in the United Republic of Tanzania

The REDD+ scheme was designed to provide countries with incentives to reduce emissions from deforestation and forest degradation and encourage them to adopt a sustainable forest management approach – and by so doing enhance their stocks of forest carbon. This report has already shown that, even without REDD+ results-based payments or finance, it makes economic and financial sense for the United Republic of Tanzania to expand the current forest estate in the interests of maximising national welfare. In recognition of the global values of the country's forests, however, international resources such as REDD+ could augment forest management and conservation at a national level through:

- Providing financial support through results-based payments or finance
- Providing technical assistance and training

- Supporting in-country efforts designed to create awareness of international agreements and treaties that encourage sustainable forestry management
- Supporting in-country advocacy efforts and efforts designed to motivate better forest policy formulation

It can be concluded that REDD+ investments would make an important contribution to a green economy transformation in the United Republic of Tanzania. They would achieve this by supporting investments that reduce household dependence on forests and increase household participation in forest management and conservation, by encouraging strategies that increase the contribution of the private and public sectors in forest conservation and management and by conducting activities that promote greater participation of the international community in supporting forest conservation and management efforts in the United Republic of Tanzania.

Notes

1 This issue is explored in greater detail in the section titled "Forest ecosystem services included in the valuation study" of this chapter.
2 Sectoral growth and price impacts, exogenous demand shocks, and exogenous price shocks were also modelled. See UNEP (2015).
3 Technically, an increase is exogenous if it comes from outside the system being modelled. To give a practical example, increased demand by the rest of the world for carbon sequestration services (or habitat protection services) provided by Tanzanian forests would present an exogenous increase for output from the forestry and hunting sector. Another way to think about it is that the Tanzanian Planning Commission could aim to increase the output from forestry and hunting by 10% over the next five years as a policy target (presumably to increase availability of goods and services to consumers from the sector).
4 A multiplier in economics is a factor of proportionality that measures how much an endogenous variable changes in response to an exogenous variable. The multiplier will not only provide information on whether the endogenous variable in question relatively increases or decreases (i.e., the direction of change), it will also provide information on the relative magnitude of the change (the size of the increase or decrease). This section will demonstrate the value of the multiplier analysis to policies affecting forestry and hunting sectors in the United Republic of Tanzania.
5 The economic notion of "elasticity", as used in the present report, refers to the degree of responsiveness of a sector to a change in other determinants and variables.
6 The issue of deconstruction, and its value to the formulation of policy, is explored further in the section titled "Deconstructing the multiplier".
7 Downstream sectors are those which use output from forestry and hunting in their own production, thereby adding value in the forestry and hunting sector.
8 The number assigned to each point on the graph corresponds to the order in which activities appear in the input matrix sheet.
9 This is similar to the multiplier product matrix – also known as first order intensity field of influence.
10 In a framework of the social accounting matrices type, one production activity can influence another through the intermediate effects on factors and institutions (households) which are considered exogenous in the input-output framework (Defourny and Thorbecke, 1984).
11 Detailed analysis and explanation of how discounted and undiscounted costs and benefits were defined is presented in section 3.10 and 3.11 of UNEP (2015).
12 Full list can be found in section 5.1 of UNEP (2015).

References

Agrawal, A.A., A.P. Hastings, M.T.J. Johnson, J.L. Maron, and J. Salminen (2012). Insect herbivores drive real-time ecological and evolutionary change in plant populations. *Science*, 338, 113–116.

Defourny, J., and E. Thorbecke (1984). Structural path analysis and multiplier decomposition within a social accounting matrix framework. *The Economic Journal*, 94, 111–136.

Division of Environment (2009). *Fourth National Report on Implementation of the Convention on Biological Diversity*. Division of Environment, United Republic of Tanzania, Dar es Salaam, 80 pp.

Division of Environment (2013). *National Strategy for Reduced Emissions from Deforestation and Forest Degradation (REDD+)*. Division of Environment, Vice-President's Office, United Republic of Tanzania, Dar es Salaam, 73 pp.

European Commission, Food and Agriculture Organization of the United Nations, International Monetary Fund, Organisation for Economic Co-operation and Development, United Nations and World Bank (2014). *System of Environmental-Economic Accounting 2012: Central Framework*. Studies in Methods, Series F, No. 109. Sales No. E.12. XVII. 12.

FAO (2010). *Global Forest Resources Assessment 2010: Country Report, United Republic of Tanzania*. Food and Agriculture Organization of the United Nations, Rome, 56 pp.

FBD (2000). *Forestry for Poverty Reduction and Economic Growth: National Forest Programme Formulation in Tanzania (Draft)*. Forestry and Beekeeping Division, United Republic of Tanzania, Dar es Salaam.

Hassan, R., and E.D. Mungatana (eds.) (2013). *Implementing Environmental Accounts: Case Studies from Eastern and Southern Africa*. Springer, The Netherlands.

Kaale, B.K. (2001). *Forest Landscape Restoration*. United Republic of Tanzania Country Report, IUCN/WWF, 53 pp.

Kahyarara, G., W. Mbowe, and O. Kimweri (2002). *Poverty and Deforestation around the Gazetted Forests of the Coastal Belt of Tanzania*. Research on Poverty Alleviation, Research Report.

Millennium Ecosystem Services Assessment (2005). *Ecosystem and Human Well-Being*. World Health Organization, Geneva.

MNRT (1998). *National Forest Policy*. Ministry of Natural Resources and Tourism, United Republic of Tanzania, Dar es Salaam, 59 pp.

MNRT (2000). *Forestry Outlook Studies in Africa (FOSA)*. Ministry of Natural Resources and Tourism, United Republic of Tanzania, Dar es Salaam, 36 pp.

NAFORMA (2014). *The National Forest Resources Monitoring and Assessment 2009–2014*. Tanzania Forest Service Agency, United Republic of Tanzania, Dar es Salaam.

National Bureau of Statistics, Ministry of Finance (2013). *National Accounts of Tanzania Mainland 2001–2012*. United Republic of Tanzania, Dar es Salaam.

Ngaga, Y.M. (1998). *Analysis of Production and Trade in Forestry Products of Tanzania*. Doctoral Scientiarium Thesis. Department of Forest Sciences, Agricultural University of Norway, Ås, Norway.

Ngaga, Y.M. (2011). *Forest Plantations and Woodlots in Tanzania*. African Forest Forum, Nairobi, 80 pp.

Parra, J.C., and Q. Wodon (2009). *SimSIP SAM: A Tool for the Analysis of Input-Output Tables and Social Accounting Matrices*, version 1.1. World Bank, Washington, DC.

Resource Economic Analysis of Catchment Forest Reserves in Tanzania (2003). Ministry of Natural Resources and Tourism, Forestry and Beekeeping Division, United Republic of Tanzania, Dar es Salaam.

Roe, D., and J. Elliot (2010). *The Earthscan Reader in Poverty and Biodiversity Conserva-tion.* Earthscan, London.SEEA EEA (2012). SEEA Experimental Ecosystem Account-ing. https://unstats.un.org/unsd/envaccounting/eea_white_cover.pdf

Tanzania National Bureau of Statistics (2012). Statistical Abstract. www.nbs.go.tz/nbstz/index.php/english/tanzania-abstract/372-statistical-abstract-2012

TEEB (2010). Mainstreaming the economics of nature: A synthesis of the approach, con-clusions and recommendations of TEEB.

United Nations Environment Programme (UNEP) (2013). *Annual Report Spotlights Action on Key Environmental Issues.*

United Nations Environment Programme (UNEP) (2014). *Building Natural Capital: How REDD+ Can Support a Green Economy.* Report of the International Resource Panel, United Nations Environment Programme, Nairobi.

United Nations Environment Programme (UNEP) (2015). *Forest Ecosystems in the Transi-tion to a Green Economy and the Role of REDD+ in the United Republic of Tanzania.* United Nations Environment Programme, Nairobi.

Village Land Act (1999). http://extwprlegs1.fao.org/docs/pdf/tan53306.pdf

VPO (1998). *Tanzania Country Study on Biological Diversity.* Vice-President's Office, United Republic of Tanzania, Dar es Salaam, 163 pp. (UNEP 2014).

10 Natural capital and GDP of the poor in Vietnam

Pavan Sukhdev and Kaavya Varma

Background: green economy and natural capital

This chapter is designed to provide an assessment of the relevance and applicability of a Green Economy – System Dynamic Modelling (GE-SDM) approach in the context of Vietnam (see Box 10.1 for the overall structure of GE-SDM). Providing options by displaying what the impacts are on poverty, environment and economy, the purpose of the approach is to assist Vietnam's national and provincial policy makers in making more informed choices about planned interventions.

Green economy 2.0

Green economy 2.0 refers to what exactly could be improved within the traditional green economy approach and what specific areas require additional attention and research. There is no clear consensus on what constitutes the basis for a green economy 2.0, bearing in mind, however, that these recent concerns lack recognised literature and clear framing. It is thus commonly acknowledged that green economy 2.0 includes important dimensions that are not properly addressed in a traditional green economy framework by providing more than just guidance.

Detailed assessments which have the potential to address natural capital issues in various developing countries including Vietnam (see the section titled "Case-study area: the Usumacinta floodplain") can be potentially covered by a green economy 2.0 approach. The traditional green economy approach emphasises poverty eradication, but does not, for example, fully incorporate the implications of the diverse features on which the poor build their vital income and well-being. A GE-SDM approach which includes an equity indicator like GDP of the Poor can significantly enhance the ability of governments to measure their progress in achieving inclusive development. Such an equity indicator can be applied either separately, in addition to conventional metrics like GDP, to improve accuracy in determining the welfare of the rural poor or it can supplement ongoing green growth initiatives. Such flexibility ought to increase the likelihood of adopting GDP of the Poor in governance processes, since the challenges of integrating new systems into existing procedures are reduced.

The green economy 2.0 approach along with extensive data collection and analysis using a set of more appropriate green economy indicators will enable the United Nations Environment Programme (UNEP) to offer (1) better evaluation

Box 10.1 Structure of GE-SDM

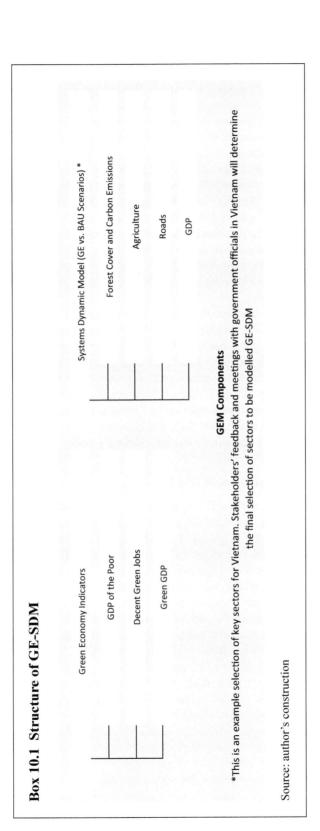

Green Economy Indicators

GDP of the Poor

Decent Green Jobs

Green GDP

Systems Dynamic Model (GE vs. BAU Scenarios) *

Forest Cover and Carbon Emissions

Agriculture

Roads

GDP

GEM Components

*This is an example selection of key sectors for Vietnam. Stakeholders' feedback and meetings with government officials in Vietnam will determine the final selection of sectors to be modelled GE-SDM

Source: author's construction

practices and (2) conclusions and implications of a GDP of the Poor measure. In this regard, green economy 2.0 proposes a more specific analysis that provides more detailed and exclusive guidelines. In other words, the green economy 2.0 approach aims to emphasise the *crucial importance* of *on-ground data collection* resulting in a less generalised, more context-based analysis that is better adapted to a specific environment.

Another gap in the traditional green economy analysis is the socio-economic exclusion of certain minority communities. A case study on Vietnam demonstrates well how agricultural minorities established in mountainous regions are excluded from the social and economic benefits of development (Trung et al., 2007). Green economy 2.0 thus emphasises the necessity of local involvement and the *focus on local fiscal management, including investments that originate from regional economies.* International investments are equally necessary for a green economy transition, but such financing may induce dependency and thus cannot replace crucial local funding.

Finally, the traditional green economy fails to deal with issues regarding *institutional coordination and transparency*. Such factors are crucial in the process of implementing a green sector. A lack of arrangement and delimitation within policy-making institutions can reduce efficiency considerably.

The green economy 2.0 approach could improve the implementation conditions of a green economy in a multitude of countries, especially because the focus of this approach originates from assessments conducted within these countries. However, it is important to note that the green economy 2.0 is not a perfectly flawless approach. There is a continued need for technical inputs.

Project for ecosystem services (ProEcoServ)

The growing demand for ways to effectively monitor and evaluate the progress of a green economy transition has triggered the enactment of more specific projects, including UNEP's Project for Ecosystem Services (ProEcoServ). ProEcoServ aims to support a better integration of ecosystem assessments into national policy-making processes and thus follows the rationale of a growing necessity for the protection of ecosystem services and biological resources.

Pilot countries (Vietnam, Chile, South Africa, and Trinidad and Tobago) have all committed to recognise the significant economic value of ecosystems. The ultimate aim of ProEcoServ is to consolidate poverty mitigation and sustainable development through economic valuation of ecosystem services. This is achieved by, for example, the development of an institutional and scientific platform to support the coalescing of decision making and ecosystem service management (ProEcoServ, 2015). A GE-SDM application in Vietnam could potentially contribute to the achievement of ProEcoServ objectives by:

- Providing an appropriate range of green economy indicators (three for simplicity and easy adoption) in conjunction with a system dynamic model that can be an effective toolkit for policy-making

- Testing and vetting an example of a GDP of the Poor implementation to develop guidance on pro-poor policies, especially at provincial government level by establishing more effective training and capacity building workshops
- Demonstrating green economy implementations in Vietnam and Indonesia, which provides justification for replication in other geographical regions

Significance of natural capital

The significant value of natural capital and its services for both consistent economic growth and for enhanced social welfare is well recognised. A number of studies and initiatives such as the Millennium Ecosystem Assessment and The Economics of Ecosystems and Biodiversity (TEEB) have highlighted the critical contributions of natural resources to human well-being as well as the poor integration of their benefits into existing valuation and policy-making frameworks. Such findings have encouraged efforts to develop more appropriate methodologies that measure and integrate the contributions of ecological services into decision-making processes and current measures of progress such as GDP. The failure of conventional GDP to reflect environmental impacts, social costs and income inequality (Costanza et al., 2014) is widely agreed upon. Initiatives such as the System for Environmental-Economic Accounting (SEEA) and Wealth Accounting and the Valuation of Ecosystem Services (WAVES) aim to correct this oversight and provide approaches to account for ecosystem services so that there is genuine growth reflected through long-term advances in environmental preservation, wealth and poverty reduction (WAVES, 2014; European Commission et al., 2014).

Indicators for green economy transitions

Building upon a range of economic valuation techniques to account for goods and services from the environment, the right indicators are needed to measure the success of their integration into development policies, particularly whether cross-sectoral environmental, economic and equity benefits are indeed attained through integration of better green accounting methodologies. Several efforts aim to construct effective sets of indicators to assist evaluation of progress. These sets of metrics range from 26 (see OECD, 2014) to 60 (see GGGI, 2012). A number of countries have developed their own set of sustainability indicators to support policy reforms and green growth. These countries include Indonesia (Gustami, 2012), Thailand (Panmanee and Sutummakid, 2012), Ghana (Bawakyillenuo, 2012) and Uruguay (Duran and Chiesa, 2012).[1]

Overall, these indicators incorporate the concept of "Green GDP" as well as skills and job creation based on sustainability. However, they would benefit from an ability to conduct micro analysis about household well-being as well as from an indication of whether national growth is improving the conditions of the poor. An inclusive indicator like "GDP of the Poor", which measures the value of household incomes of rural and forest-dependent communities including the contribution to livelihoods and well-being from biodiversity and ecosystem services (TEEB, 2008), can fill this gap and strengthen the ability of the emerging green growth approaches to deliver equity in development.

The need to assess improvements in well-being of the poor is especially relevant if countries seek transition towards a green economy. UNEP defines a "green economy" as one that results in improved human well-being and social equity, whilst significantly reducing environmental risks and ecological scarcities (UNEP, 2011). Many countries are actively reviewing their policies, fiscal incentives, industrialisation plans and status of natural resources to transition towards more inclusive growth that reduces income gaps (see PAGE, 2015).

Applicability of a GE-SDM approach in Vietnam

In Vietnam a number of efforts have been made to preserve natural resources and understand the value of their impact on the country's socio-economic well-being. These efforts include numerous legislative as well as ongoing project-level initiatives such as the Green Growth Strategy with vision to 2050, the Socio-Economic Development Strategy (2011–2020), the Strategic Orientation for Sustainable Development policy and Vietnam's development of a green Gross Domestic Product (GDP) index. Vietnam has already selected specific areas for interventions, notably reducing greenhouse gas emissions, promoting clean technologies and renewables, greening production processes, restoring natural capital, greening lifestyles and encouraging sustainable consumption. In addition to better allocation of annual budgets, streamlined institutional arrangements and innovative fiscal resource generation schemes are supposed to provide support in achieving these targets (Ministry of Planning and Investment, n.d.).

Vietnam is thus considerably advanced in its efforts to mainstream green economic development (a target for 2050) with various greening policies and programs in place. Vietnam would however still benefit from an overarching measurement approach to assess the progress in transitioning towards a green economy. Such approach would deliver the status of natural resources and thereby important information to policy makers that can support achievement of sustainable as well as economic development goals. Moreover, an inclusive equity indicator like GDP of the Poor can help the national and provincial governments to identify whether interventions being implemented are resulting in betterment of the poor.

In Indonesia three indicators, namely GDP of the Poor, Green GDP and Decent Green Jobs, are implemented through a System Dynamic Model to support national and local planners in assessing progress towards a green economy. The SDM, which is a computer modelling system, allows policy makers to compare business-as-usual versus green economy trajectories for selected sectors (e.g. energy, transport, forest) over time. Indicators are used as inputs, and the model features feedback loops and delays that mimic complications of a real economy. The approach allows planners to visibly see whether a selected intervention will result in competitive revenue generation, employment creation and poverty reduction (Sterman, 2000) in 1-, 5-, 10- and 20-year timeframes.

Customisable at a provincial level, the GE-SDM approach was launched in Central Kalimantan (Kalimantan Tengah: Kalteng) as Kalteng's Green Economy Model (KT-GEM). The model incorporates unique provincial environmental and socio-economic characteristics to assist decision makers in evaluating the success of a green economy transition and to help them create jobs and development through better management of Kalteng's natural resources.

The application of a similar GE-SDM approach in Vietnam could help the government assess the effectiveness of its measures in achieving transition to a green economy and propose where policy changes might be required in support of development goals. A GE-SDM implementation could also contribute to the development of national and provincial green economy models in Vietnam and establish processes that enable periodic assessments of the well-being of the poor at individual household levels through GDP of the Poor (see Box 10.2). Such micro-level analysis has potential to strengthen Vietnam's capacity to generate

---

Box 10.2 Expected outcomes for Vietnam based on full-scale GE-SDM implementation

Expected outputs for Vietnam based on full-scale GE-SDM implementation:

1. Assessment of the status of Vietnam's biodiversity and ecosystem services through Green GDP
2. Identification of economic sectors with the scope for additional employment creation through Decent Green Jobs
3. Assessment of the extent of nature's contribution to the incomes of rural poor households based on GDP of the Poor surveys in pilot Province
4. Based on interest development of Vietnam's Green Economy Model with the following benefits:

Review of key Green
Growth Policies

Identification of gaps
in natural capital data

Diverse and innovative   Pro-poor
livelihood opportunities
at the households level

Better transparency   Effectively supporting
in decision making   Pro-growth   Green Economy goals

Provincial level
economic revenue generation
possibilities based on improved
management of   Pro-environment
natural capital

Capacity building of local planners

Improved coordination
between ministry officials

equitable growth and meet the proposed Sustainable Development Goals (SDGs). Thirteen of the 17 SDGs are directly or indirectly affected by natural resource preservation and the cascading impacts on welfare of the poor (UN, 2015). Women and ethnic minorities are, for example, more likely to be affected due to the fact that these members of society very often are everyday consumers of natural resources e.g. they collect fuel wood for cooking, sell non-timber forest products (NTFPs), obtain water from rivers and streams for consumption or get free feed from the forests for livestock. An enhanced capacity to assess the improvement in incomes of the poor can support genuine development in Vietnam.

Vietnam's background

Status of natural resources

According to the Ministry of Agriculture and Rural Development, Vietnam's forests covered 39.7% of the total land area in 2011 (MARD, 2011) – a considerable increase from 1990 when the forest cover was 27.2% (FAO, 2010). Forests in Vietnam are categorised as production forests (6.3 million ha), protection forests (4.8 million ha) or special-use forests (2 million ha).[2] In 2005 forest economic production, including plantations, harvesting and other services, constituted 1.2% of the GDP (FAO, 2009). Deforestation remains a concern in many areas in Vietnam with habitat fragmentation being a key problem (Ngai et al., 2009).

Vietnam is ranked as the 16th most biodiversity rich country in the world while agro-biodiversity levels are among the world's highest (USAID, 2013). However, illegal wildlife trade (WWF, 2012), illegal import of timber from neighbouring countries (USAID, 2013), land grabbing of and encroachment into Protected Areas (FAO, 2010) and loss of mangroves remain prevalent issues (USAID, 2013). USAID report *Vietnam Tropical Forest and Biodiversity Assessment* (2013) lists the following sources of forest degradation in Vietnam:

*   Deforestation and land use change

    It has been estimated that primary forests, due to high deforestation rates, represent only 1% of Vietnam's total forest cover (FAO, 2010). Even though total forest cover has increased, trends show that timber plantations and cacao cultivation replace much of the mature forest. A significant amount of biodiversity resides in the primary forests, but these forests are negatively affected by the increases in commercial agriculture and by the rising demand for infrastructure that causes habitat fragmentation. Mangrove abundance has also been shown to be adversely affected by aquaculture production including large-scale clam and shrimp farms.

*   Illegal wildlife trade

    Vietnam serves as an effective intermediary and supply port for the regional demand for wildlife species. High profits, poor enforcement and a cultural preference for endangered animal and plant species continue to drive these activities.

- Illegal logging

  In addition to the primary forest cover losses and habitat fragmentations, Vietnam contributes to a thriving illegal trade of timber imported from its neighbouring regions. The production of furniture, paper and pulp has grown fivefold between 2000 and 2008, indicating that these industries will remain strong economic contenders based on the interest of the government to expand into new markets.

  (To Xuan Phuc and Canby, 2011)

- Overuse of natural resources

  Exploitation of natural resources such as forests, wildlife, mangroves and marine ecosystems is a concern in Vietnam especially in densely populated settlements around Protected Areas.

- Pollution

  Water pollution in lower watersheds of major rivers and from urban waste-water discharge needs to be addressed. Urban biodiversity is similarly important to the well-being of inhabitants in Vietnamese cities. In the city of Canberra in Australia, for example, local authorities have been planting and maintaining trees resulting in climate regulation, air pollution reductions and approximately US$4 million annually in energy cost savings.

  (TEEB, 2010)

- Infrastructure development

  Vietnam is actively constructing roads, canals, dams, dikes, ports etc. especially with the aim to reduce climate related risks. However, a number of these dikes and canals, constructed in areas vulnerable to sea level rise and storm swells, cause aquatic habitat fragmentation and disturb natural ecosystem dynamics. Hydropower projects and road constructions have similar adverse impacts on aquatic biodiversity and habitat fragmentation, respectively.

- Reduction of agro-biodiversity

  Vietnam is the second largest exporter of rice after India. While reducing food insecurity, agricultural practices have led to monocropping and focus on high-yield crop varieties.

  (Phan Truong, 2003; Sida, 2004)

Weak management of protected areas and inadequate enforcement of legislation significantly impact on natural capital in Vietnam when development-oriented initiatives are being implemented (USAID, 2013).

Socio-economic interlinkages

The economy of Vietnam experiences rapid growth. It is expected that Vietnam is the 35th largest economy in the world by 2020 with a per capita GDP of US$4,357 (Karmel, 2007). According to a study by PricewaterhouseCoopers, Vietnam is

likely to have the fastest growing economy amongst all emerging world economies by 2020, with a potential annual growth rate of approximately 10%. Such
prospects suggest that Vietnam's economy increases to 70% of the size of the UK
economy by 2040 (PWC, 2008).

Vietnam is the second largest exporter of coffee and rice in the world. The
industrial sector constituted 40.1% and the service sector 38.2% of GDP in 2004,
but the agricultural sector employs 80% of the population (ODI, 2011). Forestry
is estimated to represent only 1% of GDP, but this estimate does not account for
the contribution of forests to industrial production, firewood or a range of other
goods and services (Vietnam National Report, 2005). Thus, the economy of the
country, regardless of fast growing manufacturing and industrial sectors, is deeply
dependent on its natural capital for its long-term socio-economic well-being.

Nearly 24 million people are estimated to have lived in or around forests in Vietnam in 2002 (Forest Science Institute of Vietnam, 2009). Overall poverty levels
dropped to 20% in 2004 from 58% in 1993 (Swinkels and Turk, 2006). However,
inequality has increased and income gaps have widened. Economic growth is thus
not filtering down to the poor. Ethnic minority poverty rates in particular have
remained high in all regions except the Mekong Delta. In some regions ethnic
minority poverty rates are above 50% even as high as 70% (Swinkels and Turk,
2006). Food poverty is also considerably more critical for ethnic minorities. Agriculture and natural resources, therefore, play a more important role for such minorities that are predominantly located in rural areas (Vietnam National Report, 2005).

Policy review

Vietnam's National Green Growth Strategy (VGGS)

Vietnam's Green Growth Strategy was adopted in September 2012 (Socialist Republic of Vietnam 2012). The strategy has three main objectives: (1) ensure low-carbon
growth, (2) encourage green production systems and (3) promote green lifestyles. Targets to be achieved by 2050 include efficient resource use, reduction of environmental
degradation, reduction of greenhouse gases, restoration of natural capital and creation
of green jobs (GIZ, 2014). The specific targets of the Green Growth Strategy are:

- To reduce GHG emission intensity by 8–10% compared to the 2010 level
  by 2020
- To reduce energy consumption by 1–1.5% per year per unit of GDP
- To increase the value of high-technology and green technology to a share
  of 42–45% of GDP by 2020
- To enrich natural capital to 3–4% of GDP
- To mainstream green economic development by 2050

(National Green Growth Strategy 2012)

In the period 2011–2020, VGGS plans focus on communication, capacity development, policy mechanisms and progress evaluation including identification of relevant

indicators. In addition to the VGGS, Vietnam has introduced an Environmental Protection Tax (on fossil fuels and other environmentally harmful substances) and an
Environment Protection Fee (for wastewater, solid waste and extractive industries).[3]

Socio-Economic Development Strategy (2011–2020)

The Socio-Economic Development Strategy is targeted towards achieving a mid-
dle-income status in Vietnam in terms of per capita income of at least US$3,000
and significant improvements in human development and poverty alleviation by
2020 (ADB, n.d.). The policy also aims to support Vietnam in transforming a poor
agriculture-based economy to a wealthier market-based economy that is constantly
growing and integrated into regional and international communities.

As a part of the Socio-Economic Development Strategy, Vietnam has defined 12
development goals known as Vietnam's Development Goals (VDGs). The VDGs
are inspired by the Millennium Development Goals by including targets on reducing vulnerability, improving governance for poverty reduction, reducing ethnic
inequality and ensuring pro-poor infrastructure. The promotion of gender equality
is refined by the inclusion of two additional targets: (1) reducing the vulnerability
of women to domestic violence and (2) enhancing women's access to land and
credit (U.N. Habitat, n.d.).

To support the overall goals of socio-economic development a number of other
policies have also been implemented such as the Comprehensive Poverty Reduction and Growth Strategy in 2003 as well as a green GDP calculation.

National Strategy for Environmental Protection
until 2010 and vision towards 2020

The National Strategy for Environmental Protection focuses primarily on land
matters, water conservation, biodiversity, forest cover, and air pollution (National
Strategy for Environmental Protection, 2003). Clean production is one of the main
visions of this legislation and the strategy is critical in setting out the requirements
for environmental impact assessments, strategic environmental assessments as well
as in extending producers' responsibility in waste management (Mori et al., 2013).

Green GDP index

The Central Institute for Economic Management has developed a green GDP for
Vietnam based on the SEEA methodology for environmental accounting (Vu Xuan
Nguyet Hong et al., 2012). The index was supposed to be implemented in 2014,
but the latest status of the initiative is uncertain. The initial plan, however, was
to focus on natural resource stocks; non-renewable energy resources, including
petroleum, coal and gas; pollution accounts; and environmental public spending
accounts (CIEM, 2012) in the first phase.

Vietnam has implemented several ongoing sustainability, green economy and
equity-oriented policies and programs. However, based on the initial review of

> ## Box 10.3 Value added to Green Growth Strategy 2050 by a GE-SDM approach
>
> Value added to Green Growth Strategy 2050 by a GE-SDM approach:
>
> - Contribute to the development of socio-economic and environmental indicators through the three proposed green economy indicators i.e. GDP of the Poor, Decent Green Jobs and Green GDP
> - Evaluate the success of programs and policies implemented thus far in achieving the targets of Green Growth Strategy 2020
> - Assist policy makers in testing potential interventions by enabling them to observe results on equity, environment and economic productivity in 1-, 10- and 20-year timeframes and thus support the decision on whether to pursue business as usual or green economy trajectories
> - Support policy makers in identifying key sectors of the economy and their linkages with natural capital, so that they can determine where investments need to be directed to reduce greenhouse gases, to encourage green lifestyles and to restore natural resources
> - Calculate GDP of the Poor in a pilot province and train local officials in the process so that they can monitor the incomes of selected rural households periodically to assess whether national growth improves their well-being

strategies, initiatives and legislation, it is recommended that Vietnam implements a GE-SDM approach to measure and guide progress towards the goals of the Green Growth Strategy 2050. GE-SDM is considered as a suitable approach because the Green Growth Strategy aims to achieve cross-sectoral benefits and to provide the foundation for an equity measure, such as GDP of the Poor, to be integrated into the national strategy of the country (see Box 10.3).[4]

Recommendations for GE-SDM implementation in Vietnam

GDP of the Poor

It is recommended that at the first stage of an implementation of a GDP of the Poor metric in Vietnam, a pilot province is selected to examine data availability and the extent to which local government officials understand the basis for GE-SDM i.e. the relationship between natural resources and incomes of rural households supported by green growth policies. A study by McElwee (2008) in one of the poorest areas of Vietnam, the Ha Tinh province, shows that a number of non-timber forest products like palm leaves, medicinal plants, rattan, aromatics and wildlife are collected and sold for cash incomes by rural households. Households in Ha Tinh also

Table 10.1 Average cash income generation from forests in Ha Tinh

Forest-based activity	Total number of households collecting (% of total)	Number of households selling (% of total)	Average cash obtained by selling product (VND yr) for those households with cash income from product	Average % contribution to overall household cash income for households that sold product
Fuelwood	79 (76%)	36 (35%)	409,417	9.8
Leaves	54 (52%)	36 (35%)	365,681	8.4
Fruits	36 (35%)	8 (8%)	178,438	4.9
Timber	27 (26%)	23 (22%)	510,870	12.1
Rattan/bamboos	27 (26%)	5 (5%)	184000	4.2
Charcoal	19 (18%)	19 (18%)	1,173 947	25.0
Medicinals	14 (13%)	7 (7%)	232 429	5.2
Fodder	8 (8%)	0	0	0
Honey	5 (5%)	3 (3%)	733,333	11.5
Aromatics/oils/others	4 (4%)	1 (1%)	20,000	<1
Animals	3 (3%)	1 (1%)	500,000	13.1

Source: McElwee (2008)

depend on fuelwood (collected by 76% of the households and sold by 35%) and edible plants and fruits. A total of 88% of the households in Ha Tinh harvest forest resources for both commercial and subsistence uses demonstrating a significant dependence on nature amongst the poor (ibid).

In an upland forest-dwelling community in northern Vietnam, Quang and Nor-iko (2008) find that more than 65 percent of the annual income of poor households is derived from forests, whereas the same measure is less than 40=% for richer households. For some households, forest-derived income constitutes more than 75% of total household income (Quang and Noriko, 2008). Timber, non-timber forest products and agricultural crops cultivated on forestland make up the greatest proportion of the incomes of the poor. In fact, the study finds that the very poor in particular are the ones whose livelihoods are most heavily dependent on forests (ibid).

Overall, there are a significant number of communities that depend on forest resources in the rural regions of Vietnam. Fifty-three ethnic minorities in Vietnam depend mainly on forests for their livelihoods. While they constitute only 13% of the population, they account for 30% of the poor in Vietnam (World Bank, 2008). The ability for policy makers to assess whether economic development is improving the welfare of these households is critical for poverty alleviation. Such ability can also contribute to enhanced food security planning and stability in agriculture as these communities generally have small plots of low quality land (ibid).

Methodology and proposed pilot

A GDP of the Poor survey should be executed subsequent to receiving in-depth stakeholder feedback on the socio-economic factors of the selected pilot region. Previous projects that have collected household-level data to examine the availability of relevant statistics will be reviewed and applied to support pilot region implementation. UNEP's ProEcoServ, for example, has been working in the Ca Mau region to study the status of particularly mangroves and to capture the intense pressures on ecosystem services from land-use change, population growth and unsustainable use (Kumar, 2013). The purpose of ProEcoServ in Vietnam is to improve the integration of ecosystem services and vulnerabilities due to anthropogenic activities as well as climate related changes into the Green Growth Strategy, the National Strategy for Environmental Protection and Party Resolution 24-NQ/TW on Active Response to Climate Change, Improvement of Natural Resource Management and Environmental Protection (ibid). Ecological data collected by ProEcoServ and the Institute of Strategy and Policy on Natural Resources and Environment (ISPONRE) in Ca Mau can potentially serve as input into a GDP of the Poor survey (if stakeholders find Ca Mau suitable as pilot region). Survey forms would thus include key forest resources of the region and therefore be better suited to obtain relevant information from households. In addition, existing partnerships between villagers and local experts, formed by ProEcoServ, could facilitate the establishment of appropriate questions to be included in the survey.

ProEcoServ can also support policy making at the national level by providing recommendations on how to enhance the welfare of poor households more effectively through better management of critical natural resources. The value of the services provided by mangrove ecosystems to the regional economy in terms of reducing adverse climate impacts and thus, contributing to resilience in vulnerable coastal villages, further strengthens the argument for mangrove protection and highlights its priority in national climate risks mitigation strategies.

A number of surveys on rural poverty in Vietnam are conducted periodically by NGOs, for example the Vietnam Household Living Standards Surveys and the Participatory Poverty Assessments (World Bank, 2012). Most of these surveys examine per capita consumption, calories absorbed, types of food eaten and non-food spending, but such data can be analysed to provide contextual information in support of a GDP of the Poor survey. It would for example be relevant to examine whether the popularity of some forest-related natural resources has diminished or increased over time and also whether rural poverty rates have declined (see Table 10.2). Such information will enable local policy makers to target scarce funds for investments in protection of natural resources towards the resources that are critical for the poor and that demonstrate trends of degradation.

The following methodology will be used to implement GDP of the Poor in Vietnam (see Box 10.4).

Table 10.2 Existing rural and urban poverty estimates for Vietnam to inform baseline development

	GSO-WB Poverty Rate		Official Poverty Rate	
	Incidence (%)	*Contribution to total (%)*	*Incidence (%)*	*Contribution to total (%)*
All Viet Nam (national)	20.7	100	14.2	100
Urban	6.0	9	6.9	14
Rural	27.0	91	17.4	86
Red River Delta (Hanoi)	11.4	12	8.4	13
East Northern Mountains	37.7	21	24.2	20
West Northern Mountains	60.1	9	39.4	9
North Central Coast	28.4	16	24.0	20
South Central Coast	18.1	7	16.9	10
Central Highlands	32.8	10	22.2	9
Southeast (HCMC)	8.6	7	3.4	4
Mekong Delta	18.7	17	12.6	17

Source: GSO-WB poverty line, developed jointly by the General Statistics Office and the World Bank in the mid-1990s (World Bank, 2012)

Box 10.4 GDP of the Poor implementation guide

Step 1 Village selection is drawn in appropriate proportion to the total number of rural villages in the province.

Step 2 The types of villages are identified based on the provincial context. For example, in Central Kalimantan the village categories were forest, riverside, rural mixed with rattan and rural mixed with coal.

Step 3 A survey questionnaire is developed to elicit information about sources of cash and non-cash incomes per household.

Step 4 Sample households are selected from each category of village.

Step 5 Survey team members are selected based on previous experience with surveying, data gathering and familiarity with areas in pilot province.

Step 6 Before entering the field, survey team members are briefed about the concept of a Green Economy and introduced to the "GDP of the Poor" indicator including how it seeks to determine ecosystem service dependence.

Step 7 Teams of two are dispatched to different households in different villages to gather data simultaneously.

Step 8 A senior economist, familiar with the provincial context, oversees the data collection process and assimilates the gathered data.

Step 9 Responses from each household are noted in the Survey Form.

Step 10 Data from all households are entered into a spreadsheet to be analysed.

Outcomes: Equity indicator availability, with pilot demonstration, for integration into National Green Growth Strategy to meet goals of inclusive development. Panel data, which local officials can refer to over time, is created to determine the impacts on GDP of the Poor based on the policies they put in place.

Decent Green Jobs

The Vietnam Employment Trends Report (2007), prepared by the Labour Market Information Centre of the Bureau of Employment (BoE) with support from the International Labour Organisation (ILO), states that agriculture remains the most important economic sector in Vietnam, and informal labourers comprise 82% of all employment (Quoc, 2012). The dimensions of the informal sector as well as the associated risks, has brought attention to "decent jobs". A decent job is defined by ILO as a job that provides social rights and empowers employees. It is, however, widely recognised that there is an urgent need for an assessment of the potential that different sectors possess for Decent Green Jobs creation in order to meet the targets of the Green Growth Strategy (Quoc, 2012).

The Vietnam Employment Strategy refers to "Green Jobs" directly in section 2.1.7; however, it does not define the exact requirements in the context of Vietnam.

> "*Green Jobs*" for development and environmental sustainability
> A move towards "*green jobs*" will contribute towards a low-carbon economy, use of clean and renewable energy, resource efficiency, conservation of forests, reduced environmental pollution, enhanced solid waste management and pollutant management and a mitigated impact of industrial development on climate change"
>
> These measures and the adaptation to climate change itself by the locations, economic sectors and social groups most affected, have far reaching implications for production and consumption patterns and, therefore, for employment, incomes, and poverty patterns. These long-term transformations are likely to alter the structure of employment towards sustainability, create new jobs, to reduce labour surplus and, more in general, to positively affect the world of work on several aspects, from health and safety improvements to work organization and production patterns and value chains. Therefore, this strategy requires broad consensus among institutional actors at the province and district levels, and social dialogue mechanisms will have to ensure harmonious industrial relations for developing strategies for such transformations that are both efficient and equitable. These long-term transformations will entail transitions towards "*green jobs*" and greener enterprises.
>
> (Socialist Republic of Vietnam, 2012)

Policies in Vietnam thus reflect the need for a green jobs assessment in relevant sectors. It is recommended that Decent Green Jobs are estimated in partnership with the ILO and supplemented by existing analyses on employment trends and decency. The ILO in Vietnam and the BoE are also important actors to involve in consultation to calculate Decent Green Jobs and to determine whether national trends or in-depth regional analyses are more beneficial. The following methodology will be applied to calculate Decent Green Jobs (Box 10.5).

Box 10.5 Decent Green Jobs implementation guide

Decent Green Jobs are defined by the International Labour Organisation (ILO) as direct employment created in different sectors of the economy through activities that reduce the sectors' environmental impacts, and ultimately bring the ecological damage down to sustainable levels. Jobs are only decent if they meet social empowerment criteria such as social security for employees, opportunities to voice opinions, and so on (ILO, 2013).

Step 1 Identify international and national economists specialised in System of National Accounts, labour statistics and Green Economy approaches and frameworks.

Step 2 Identify a national partner from the National Planning Ministry, Bureau of Employment to facilitate data collection and stakeholder consultations. The national partner should host stakeholder discussions.

Step 3 Review available data: Review employment surveys (obtain access to statistical agency micro-data) and regional GDP data including the sampling methodology, to get an initial idea of the economy and employment structures to identify key sectors. This is a 1% sample.

Step 4 Identify key and relevant green sub-sectors and activities within these green sub-sectors through a dialogue-based approach. Engage other relevant ministries, experts and representatives from employers' organisations from the selected sectors. Invite these stakeholders to consultations to determine economic activities based on national law and instructions, government regulations, voluntary standards and activity-based approaches in order to identify green sub sectors. In Indonesia, for example, nine green sub-sectors were identified at the national level. The sectors within any province would be nine or less (stakeholder discussion groups should be a maximum of 10–15 people).

Step 5 Gather data /reports shared in stakeholder consultations.

Step 6 Senior Economists match regulations with International Standard Classification of Industry (ISCI).

Step 7 Determine the proportion of green activities. Review literature on individual economic activities and gather further data based on surveys, interviews, etc. to provide rationale for the proportion that is considered green.

Step 8 Validate green sectors selected by stakeholder groups and identify green sub-sectors.

Step 9 Generate employment estimates.

Step 10 Engage with "social partners" of the economy to discuss employment conditions within the identified green sub-sectors using decent work indicators. Social partners are from the employers' organisations, workers' organisations, producers' organisations and the government. The number of stakeholder consultations will be based on the number of green sub-sectors identified in the province.

Step 11 Gather data /reports shared in stakeholder consultations.

Step 12 Senior economists apply the decent work criteria to the employment estimates.

Step 13 Validate results through stakeholder consultation with social partners.

Step 14 Final validation for all nine green subsectors with stakeholders from social partners group and broader groups (experts with sector knowledge).

Outcomes: Existing sectors with greening potential identified.

Through this indicator policy makers have information regarding which sectors to invest to ensure sustainable growth and additional employment.

Green GDP

The Central Institute for Economic Management has been developing a Green GDP Index for Vietnam. It was expected that the measure would be ready to integrate into national accounts by 2014 (CIEM, 2012). However, instead of replicating Green GDP calculations, it is suggested that results from CIEM reports be applied and incorporated into GE-SDM to compare business as usual versus green economy for 1-, 5- and 10-year timeframes. In Indonesia for example, I-GEM demonstrates that, apart from the year succeeding implementation, Green GDP is consistently higher than GDP in Central Kalimantan. I-GEM also shows that incomes of rural households increase when policies account properly for the significant value of the services provided by ecosystems.

Scenarios development

The implementation of a GE-SDM approach involves the development of business as usual and green economy scenarios in different sectors. Business as usual represents following a conventional path of development based on historical data, whereas a green economy scenario implies adopting green economy interventions. Key sectors with potential for inclusive growth and improved environmental practices will be identified in collaboration with officials and key stakeholders. Possible sectors include for example agriculture, forests, transport and energy.

Two main scenarios are analysed under GE-SDM.[5]

- **Business as Usual (BAU)** case that assumes continuation of historical and present trends. This includes all policies and interventions currently active and enforced, but excludes policies planned but not yet implemented.
- **Green Economy (GE)** scenario that simulates, among others, additional interventions that promote sustainable agriculture, reduce deforestation, improve the management of fish stocks, reduce energy demand and encourage renewable energy development, cut emissions across key sectors including transport. These interventions can be tested and analysed in isolation (i.e. one by one) or simultaneously, to assess their short, medium and long term impacts across social, economic and environmental indicators.

Depending on the interest from other ministries, the results of the indicators, presented through the scenarios, could also be utilised to inform budget allocations and investments in REDD+ projects or other payment mechanisms such as Payments for Ecosystem Services. The distribution of fiscal resources would be based upon apparent benefits to poor households and environmental conservation and would, therefore, support the objectives of the Green Growth Strategy more effectively.

Efforts to model the impacts of removing fossil fuel subsidies in Vietnam versus instituting a fossil fuel tax have been undertaken using CGE (Computer General Equilibrium) modelling (Willenbocke and Cong Hoa, 2011; Minh Bao and Sawdon, 2011). An I-O analysis (Inflationary Impacts of Fossil Fuel Price Reform in Vietnam – A Static Analysis) has also been conducted to understand policy-induced inflationary impacts that are not covered by the CGE analysis in Vietnam (Bassi, 2012). However, the modelling thus far has been static and it would, therefore, be useful to develop dynamic models of selected sectors. The GE-SDM approach could apply and review data from the I-O analysis, and scenarios could be developed to add other sectors, particularly natural capital.

Data requirements and institutional partners

Data requirements and data availability for the three indicators

Data and statistics on the status of natural resources, poverty and mainstream economic sectors are available, but much of this data is insufficient to measure green growth progress and to examine linkages between environment and poverty and between environment and sector productivity. It is expected that detailed information on Green GDP is available, while GDP of the Poor and Decent Green Jobs require primary data collection. An indication of GE-SDM data requirements are summarised and listed in Table 10.3 along with potential data sources. The list is likely to change and should thus be considered provisional only.

Table 10.3 Data requirements to estimate the three indicators (GDP of the Poor, Green GDP and Decent Green Jobs)

Indicators	Broad headings of possible data requirements
GDP of the Poor	GDP of the Poor heavily depends on data from primary survey Data of forest dependent rural areas regarding – Monthly cash and non-cash income – Monthly expense of various components – Household dependency of nutrition, medicine, fuel etc. on forest – Monthly consumption of vegetable, cereal, meat, fish, eggs, milks etc. and the proportion derived from the marketplace as well as from the surrounding environment
Green GDP* * Data pertaining to Green GDP will be utilized from the Green GDP index developed by the Central Institute for Economic Management	Ecosystem System Service valuation – Data regarding services to be valued. For example, types of forest in the region, carbon stock per hectare for different types of forest, deforestation rate, regeneration rate and so on for calculating the carbon stock of an area Integrating the values of ecosystem valuation into National accounts – Green GDP = GDP(traditional) – Depletion of natural resources (change in stock) – cost of pollution – Stock of natural resources needs to be computed. – It can include stock of forest, freshwater, subsoil assets (physical and monetary terms) – Stocks to be found by (Opening stock – changes due to economic activities and other changes = Closing stock)
Decent Green Jobs	– Identification of green sectors, subsectors and activities within subsectors – Verify the identified subsectors, activities via stakeholder consultation – Stakeholder consultation should also act as a data gathering platform – Economist to apply decent work criteria to employment estimates – Validation of results through stakeholder meetings

Source: author's computation

Governmental departments, organisations and institutions to be contacted for data collection

- General Statistics Office of Vietnam
- Central Institute for Economic Management
- Department of Hydrology & Meteorology and Climate Change
- Ministry of Agriculture & Rural Development
- Ministry of Natural Resources and Environment
- Ministry of Marine Products
- Vietnam National Administration of Tourism
- The Ministry of Planning and Investment
- The information centre of the World Bank office in Vietnam

- Ministry of Labour, War Invalids and Social Affairs (MOLISA)
- Centre of the Bureau of Employment
- Labour Market Information Centre of the Bureau of Employment (BoE)
- International Labour Organisation Country Office Vietnam
- CIFOR country office
- CARE country office
- Asian Development Bank (country office)
- World Agroforestry Centre (ICRAF)
- Institute of Strategy and Policy on Natural Resources and Environment (ISPONRE)
- Universities working on relevant topics such as Vietnam National University of Agriculture, Ho Chi Minh City University of Natural Sciences, Ho Chi Minh City University of Social Sciences and Humanities, Water Resources University
- NGOs working for poverty alleviation, forest conservation, labour rights such as Vietnam Conservation Center, The Centre for Gender, Family and Environment in Development

Key institutional partnerships

To get buy-in from policy makers in Vietnam for GE-SDM and for the development of Vietnam's Green Economy Model, it is important to form partnerships with key ministries. Since the Green Growth Strategy is implemented under the guidance of the Inter-ministerial Coordinating Board, the Ministry of Planning and Investment, with its minister as the vice head of board, it is critical to get their interest in GE-SDM and GDP of the Poor. The Ministry of Natural Resources and Environment, Ministry of Finance, Ministry of Agriculture and Rural Development and Ministry of Industry and Trade are also a part of the Coordinating Board. Thus, while these other ministries are important based on the relevance of the results of the indicators for them, the Ministry of Planning and Investment remains the most critical to get buy-in from as the supporting office of the national Green Growth Strategy is physically based in its premises as well.

In terms of local partnerships, these will be based on the province selected for GDP of the Poor implementation, as that province's DoPI (Provincial Department of Planning and Investment) officials will need be introduced to green economy concepts, the linkages between long-term growth and better management of natural resources and the potential for more successful poverty alleviation in their province through GDP of the Poor assessments. However, before the DoPI officials are involved, the administration of the province needs to support the assessment within their jurisdiction and this means that the chairman needs to endorse the GE-SDM assessment first in the selected province.

Scientific institutions and experts from these institutions like the Center for International Forestry Research, the National REDD+ Network, Vietnam Forest Administration, World Agroforestry Centre and the Research Centre for Forest Ecology and Environment should be included in stakeholder consultations.

International development partners like the Asian Development Bank, the World Bank, German Organisation for International Cooperation and the Japan International Cooperation Agency (JICA) are already active in Vietnam and can provide a background of their work in natural resources and green growth so that the GE-SDM approach can build on these projects.

Notes

1 See Annex 1 for the specific set of indicators under each initiative.
2 MARD (Ministry of Agriculture and Rural Development). 2011. Decision 2089/QD-BNN-TCLN.
3 GIZ, 2011. *Environmental Taxation in Viet Nam*. Retrieved from: www.giz.de/expertise/downloads/giz2011-en-factsheet-efr-vietnam.pdf
4 Note: Recommendation is based on literature review and can be modified based on stakeholder consultations with key policy makers in Vietnam to identify other relevant policies that require support in delivering green economy goals.
5 See Sukhdev et al. (2014) for the I-GEM Full Project Report that discusses in detail the results of the scenarios for Indonesia.

References

ADB (n.d.) Viet Nam by the Numbers. https://data.adb.org/dashboard/viet-nam-numbers

Bassi, A.M. (2012) "Review of the Methodologies and Tools Employed to Analyse Fossil Fuel Fiscal Policies in Vietnam". KnowlEdgeSrl.

Bawakyillenuo, S. (2012) "Policy Context for Ghana's Activities on Green Economy Indicators, Institute of Statistical, Social and Economic Research, University of Ghana: Measuring the Future We Want". An International Conference on Indicators for Inclusive Green Growth Policies/Green Economy. United Nations Environment Programme, Centre International de Conferences Geneve, December.

CIEM (2012) Sustainable Development with Green GDP. Interview with Vu Xuan Nguyet Hong, Deputy Director, Central Institute for Economic Management.

Costanza, R., Kubiszewski, I., Giovannini, E., Lovins, H., McGlade, J., Pickett, K.E., Ragnarsdottir, K.V., Roberts, D., De Vogli, R., and Wilkinson, R. (2014) "Development: Time to Leave GDP Behind". *Nature* 505: 283–285.

Duran, V., and Chiesa, A.V. (2012) "Uruguay: Green Economy Indicators, Ministry of Agriculture, Livestock and Fishing and Ministry of Housing, Land Planning and Environment: Measuring the Future We Want". An International Conference on Indicators for Inclusive Green Growth Policies/Green Economy. United Nations Environment Programme, Centre International de Conferences Geneve, December.

European Commission, Food and Agriculture Organization of the United Nations, International Monetary Fund, Organisation for Economic Co-operation and Development, United Nations and World Bank (2014). System of Environmental-Economic Accounting 2012: Central Framework. Studies in Methods, Series F, No. 109. Sales No. E.12. XVII. 12.

FAO (2009) Vietnam Forestry Outlook Study. www.fao.org/docrep/014/am254e/am254e00.pdf

FAO (2010) Global Forest Resources Assessment 2010: Country Report, Vietnam, Rome.

Forest Science Institute of Vietnam (2009) "Vietnam Forestry Outlook Study, Asia-Pacific Forestry Sector Outlook Study II". Working Paper Series.

GIZ (2011) Environmental Taxation in Vietnam. www.giz.de/expertise/downloads/giz2011-en-factsheet-efr-vietnam.pdf

GIZ (2014) Vietnam. www.greenfiscalpolicy.org/countries/vietnam/

Global Green Growth Institute (2012) "GG Indicators, GGGI's Approach: Indicators for Green Growth Planning: Measuring the Future We Want". An International Conference on Indicators for Inclusive Green Growth Policies/Green Economy. United Nations Environment Programme, Centre International de Conferences Geneve, December.

Gustami, Z. (2012) "Indonesian Experience in Developing Sustainable Development Indicator, Ministry of Environment: Measuring the Future We Want". An International Conference on Indicators for Inclusive Green Growth Policies/Green Economy. United Nations Environment Programme, Centre International de Conferences Geneve, December.

ILO (2013) Green Jobs Mapping Study in Indonesia: Advance Draft, ILO, Indonesia Country Office, Jakarta.

Karmel, R.S. (2007) The Vietnamese Stock Market. www.fwa.org/pdf/Vietnam_posttrip_article.pdf

Kumar, P. (2013) "Tools and Models in Project for Ecosystem Services (Proecoserv)". Workshop on SEEA-Experimental Ecosystem Accounts, UN Stat, 18 Nov, New York.

MARD (Ministry of Agriculture and Rural Development) (2011) Decision 2089/QD-BNN-TCLN.

Minh Bao and Sawdon (2011) Environmental Assessment of the Potential Effects and Impacts of Removal of Fossil Fuel Subsidies and of Fuel Taxes. www.undp.org/content/dam/vietnam/docs/Publications/VN%20GHG%20envirnmnt%20analysis%20fossil%20fuel%20subsidy%20tax%20-%20paper3%20-%20final.pdf

McElwee, P. (2008) "Forest Environmental Income in Vietnam: Household Socioeconomic Factors Influencing Forest Use". *Environmental Conservation* 35(2): 147–159.

Ministry of Planning and Investment (n.d.) "Vietnam Green Growth Strategy". presentation.

Mori, A., Ekins, P., Speck, S., Lee, G.-C., and Ueta, K. (2013) *The Green Fiscal Mechanism and Reform for Low Carbon Development: East Asia and Europe.* Abingdon, UK: Routledge.

National Strategy for Environmental Protection until 2010 and vision toward 2020 (2003) Viet Nam. https://theredddesk.org/sites/default/files/national_env_strategy_1.pdf

Ngai, N.B., Tan, N.Q., Sunderlin, W.D., and Yasmi, Y. (2009) *Forestry and Poverty Data in Vietnam: Status, Gaps, and Potential Uses.* RECOFTC, Rights and Resources Initiative, Vietnam Forestry University, Bangkok.

ODI (2011) *Vietnam's Progress on Economic Growth and Poverty Reduction: Impressive Improvements.* ODI Publications, London.

OECD (2014) "The OECD Set of Green Growth Indicators". In *Green Growth Indicators 2014.* OECD Publishing, Paris.

PAGE (2015) Partnership for Action on Green Economy Brochure. www.unep.org/greeneconomy/Portals/88/PAGE/PAGE_Brochureupdate1_2015_HighRes_SinglePages2.pdf

Panmanee, N., and Sutummakid, N. (2012) "Sufficiency Economy towards Green Economy/Green Growth Policy in Thailand, Ministry of Natural Resources and Environment and Thammasat University: Measuring the Future We Want". An International Conference on Indicators for Inclusive Green Growth Policies/Green Economy. United Nations Environment Programme, Centre International de Conferences Geneve, December.

Phan Truong (2003) *In MARD & VEA, 4th Country Report: Vietnam's Implementation of the Biodiversity Convention.* Hanoi, 2008. https://www.cbd.int/doc/world/vn/vn-nr-04-en.doc

ProEcoServ (2015) ProEcoServ Project Outputs. www.proecoserv.org/proecoserv/proecoserv-project-outputs.html

PWC (2008) "China to Overtake US by 2025, But Vietnam May Be Fastest Growing of Emerging Economies". PricewaterhouseCoopers.

Quang, N.V., and Noriko, S. (2008) "Forest Allocation Policy and Level of Forest Dependency of Economic Household Groups: A Case Study in Northern Central Vietnam". *Small-Scale Forestry* 7: 49–66.

Quoc, N.C. (2012) "Greening Doi Moi: An Outlook on the Potential of Green Jobs in Vietnam". Freidrich Ebert Stiftung.

Sida (2004) Integration of Biodiversity Aspects in Development Cooperation: A Case Study. Mountain Rural Development, Vietnam. www.sida.se/contentassets/c26dbe431e2e49ef8cce22ab346a928d/mountain-rural-development-vietnam.-integration-biodiversity-aspects-in-development-cooperation--a-case-study_1301.pdf

Socialist Republic of Vietnam (2012) National Green Growth Strategy (Prime Minister Decision 1393/QĐ-TTg of 25 September 2012).

Sterman, J. D. (2000) *Business Dynamics: Systems Thinking and Modeling for a Complex World.* Irwin/McGraw Hill, Boston.

Sukhdev, P., Bassi, A., Varma, K., and Mumbunan, S. (2014) "Indonesia Green Economy Model (I-GEM)". Full Project Report. Prepared under the Low Emissions Capacity Building project, UNDP, Jakarta, Indonesia.

Swinkels, R. and Turk, C. (2006) "Explaining Ethnic Minority Poverty in Vietnam: A Summary of Recent Trends and Current Challenges". Draft Background paper for CEM/ MPI meeting on Ethnic Minority Poverty, Hanoi.

TEEB (2008) The Economics of Ecosystems and Biodiversity Interim Report, Printed by Welzel+Hardt, Wesseling, Germany.

TEEB (2010) The Economics of Ecosystems and Biodiversity: Mainstreaming the Economics of Nature: A Synthesis of the Approach, Conclusions and Recommendations of TEEB. www.biodiversity.ru/programs/international/teeb/materials_teeb/TEEB_Synth Report_English.pdf

To Xuan Phuc and Kirstin Canby (2011) *Vietnam: Overview of Forest Governance and Trade.* Forest Trends for FLEGT Asia Regional Programme.

Trung et al. (2007) "Application of GIS in Land-Use Planning: A Case Study in the Coastal Mekong Delta of Vietnam". *International Journal of Geoinformatics.* http://agris.fao.org/agris-search/search.do?recordID=NL2012004581

UNEP (2011) Towards a Green Economy, Pathways to Sustainable Development and Poverty Eradication. www.unep.org/greeneconomy

UNEP (2015) Indicators for Green Economy Policymaking: A Synthesis Report of Studies in Ghana, Mauritius, Uruguay.

U.N. Habitat (n.d.) Localising the Millennium Development Goals. http://mirror.unhabitat.org/content.asp?cid=7874&catid=71&typeid=13United Nations (2015) Sustainable Development Goals. www.un.org/sustainabledevelopment/sustainable-development-goals/

USAID (2013) "Vietnam Tropical Forest and Biodiversity Assessment". Prepared by Sun Mountain International and the Cadmus Group, Inc.

Vietnam Employment Trends (2007) Ministry of Labour, Invalids and Social Affairs, Vietnam.

Vietnam National Report to UN Forum on Forests (2005) Prepared by Department of Forestry, Ministry of Agriculture and Rural Development. www.un.org/esa/forests/pdf/national_reports/unff5/vietnam.pdf

Vu Xuan Nguyet Hong, Nguyen Manh Hai, Bui Trinh, Ho Cong Hoa, Nguyen Viet Phong, Duong Manh Hung (2012) "Green GDP index: Research for Methodology Framework Development". Final Report Submitted to British Embassy in Vietnam. Hanoi: Central Institute for Economic Management (CIEM). www.ciem.org.vn/Portals/1/CIEM/TinTuc/2012/Vietnam_Green_GDPCIEM2.pdf

WAVES (2014) The Global Partnership on Wealth Accounting and the Valuation of Ecosystem Services. www.wavespartnership.org/sites/waves/files/documents/WAVES%20Brochure-web.pdf

World Bank (2008) Report No.: 44575-VN. Washington, DC.

World Bank (2012) Vietnam Poverty Assessment. www.worldbank.org/content/./vn_PA2012Executive_summary_EN.pdf

Willenbocke and Cong Hoa (2011) Fossil Fuel Prices and Taxes: Effects on Economic Development and Income Distribution in Vietnam. www.vn.undp.org/content/dam/vietnam/docs/Publications/VN%20CGE%20analysis%20fossil%20fuel%20subsidy%20tax%20-%20paper2%20-%20final.pdf

WWF (2012) *Wildlife Crime Scorecard*. WWF, Gland, Switzerland.

11 Natural capital and the rate of discount

Anil Markandya

Introduction

What is discounting?

In order to compare economic effects that occur at different points in time, the practice of applying a discount rate to future effects has been developed. For the following reasons, individuals acting on their own as well as societies acting collectively prefer to have something now, rather than to have the same thing in the future. The difference between the value of a dollar today and a dollar in one year's time to an individual is referred to as that individual's or society's discount rate.

The value attached today to receiving one dollar a year from now is expressed as: $\frac{1}{(1+d)}$, where d is the discount rate. If d were 0.05 (5%), the value of a dollar in one year's time would be 95 cents. If the discount rate is constant, and one wants to know the value of one dollar two years from now, the 95 cents would decline by another 5% in the second year and become 91 cents. The mathematical expression for that could be written as: $\frac{1}{(1+0.05)^2}$. Extending this over a number of years would result in a value that declines geometrically.[1]

Hence, if an individual were to invest one dollar today, she would need to obtain a benefit of at least $1.05 in one year's time to consider the investment worthwhile. Likewise, the benefit required in two years would be $(1.05)^2$, equivalent to $1.1025. In T years the amount required to make the investment worthwhile would need to be $(1+0.05)^T$.

In practice, the returns to an investment in physical or natural capital accrue over several years, in which case the comparison has to be made between the investment now and the sum of these returns over the future, each discounted by the year in which it occurs. This sum is referred to as the present value (PV) and is written as:

$$PV = \sum_{t=1}^{t+T} \frac{B_t}{(1+d)^t}$$

Where B_t is the benefit in year t, in monetary terms.

The present value of the benefits of an investment can be compared with the costs, to give the net present value of the benefit associated with the investment (NPV). A positive net present value greater than the cost indicates the investment is worthwhile; a negative figure indicates it is not.[2]

When considering the flow of returns (benefits) from an investment in protecting or enhancing natural capital, it is important to remember that oftentimes the benefits are often not observed immediately after the investment is made and instead accrue over a long period of time. An example is afforestation, where planting a tree may yield no benefits for the first few years, relatively small benefits for several years afterwards and a large benefit when the tree can be logged, perhaps 50 years from now. If the benefits in that year are discounted back to today and if the discount rate was 5%, one dollar of value in year 50 would be valued as only 8.7 cents today. Furthermore, it would be worth much less (only 0.1 cent) if the discount rate was 10%. Thus, the higher the discount rate, the lower the present value will be of investments with returns far into the future.

Effects of discounting on investment in natural capital

The implications of different rates of discount on evaluating different investments in natural capital are best portrayed by examples. The following are typical of benefits relative to costs that may occur in practice, though they are not actual examples of investments.

1   An investment of $1 million to afforest an area where the monetary benefits will be realised in 50 years and no interim benefits will arise.
2   The same investment as the previous one but some benefits are obtained in years 2–49, in the form of erosion prevention and carbon sequestration. These benefits do not, however, generate a monetary return.
3   An investment of $1 million now, which will start to generate a number of benefits in 40 years' time, and these benefits will continue at a rising rate for the next 50 years and beyond. This is the broad structure of the kind of benefits associated with investments to limit climate impacts through reducing greenhouse gases.
4   An investment to protect an area of natural beauty, which generates benefits through recreational use starting after a short while and continuing over the next 50 years and beyond. Future benefits depend, however, on future recreational use and that depends on increases in living standards in the region.

These examples are discussed in terms of how the discount rate choice may influence the evaluation. The detailed calculations are not presented for reasons of space but are available on request from the author.

Afforestation with logging benefits only

In this case, if the logging generates a benefit of $5 million in 50 years' time the *NPV* is negative at a discount rate of 5%. It is only when the discount rate falls to 3.3% that the project would be deemed worthwhile. This is not untypical of forestry projects where certain species have a long maturity time and require discount rates of 3.3% or less to justify investment in them. Note, however, that the calculation of the investment has paid no attention to other interim benefits. These are considered in the next example.

Afforestation with logging benefits and environmental benefits

Suppose now that from year 2 onwards, the investment provides erosion protection and carbon sequestration services that grow with the tree, peaking in year 50. If these environmental benefits can be converted into monetary terms (there is extensive literature on how this may be done) and if they start as low as, say $10,000 in year 2, but build up to about $300,000 by year 50, then the *NPV* is positive at a discount rate of 5%. Indeed, the project would be considered as justifiable at a discount rate of up to 6.6%. The greater these environmental benefits, the higher the discount rate that results in a positive evaluation of such a project.

Investments with delayed and very long period for benefits

Consider a case where society has to make an investment of $1 million every year from now to 2050 but benefits only start to emerge in 2060. After that year, however, they grow continuously to 2100 and beyond. In 2060 the benefits are also $1 million but they increase every year after that at 5%. The result is an *NPV* that is negative at a discount rate of 5%. It requires a discount rate of less than 2.4% for this kind of program to be justified. This is a simplified example of the issue of discounting with climate change, which is discussed further in the section titled "Applications to valuing different forms of capital" . Investment in mitigation is needed in ensuing decades to prevent damages to natural and physical capital in the future (mainly from 2060 onwards). A discount rate of 5% makes such an investment questionable, and it needs lower rates to justify it.

The example raises two issues concerning discount rates. The first is that actions relating to mitigation also have major co-benefits, which occur in the short term. These take the form of lower levels of local air pollutants as fossil fuel combustion is reduced and increased resource efficiency, implying greater conservation of natural capital among others. If such benefits are included, the *NPV* of programmes to reduce GHGs can become positive at higher rates. For example, with co-benefits equal to 25% of the investments in mitigation, the required discount rate at which the programme is justified goes up from 2.4% to 3.3%.

The second issue is the great uncertainty surrounding future damages if mitigation does not take place to reduce the expected levels of climate change. The discount rate choice does not capture this uncertainty, which has been raised as a concern by some researchers. This concern is revisited in the discussion in the following section.

Investments with growing benefits over a very long period

In the fourth example, a protected area requires an investment of $1 million spread equally over the first four years (2015–2018) and starts to generate recreational benefits in year 5. These benefits are small – around $40,000 per year but continue for an indefinite period. If one considers a horizon of 2100,[3] the *NPV* for the project is negative with a discount rate of 5%. It needs a rate of less than 2.7% to be justified. Such a calculation is flawed, however, because it does not take into account the growth in recreational benefits (as populations increase and as people's disposable income and leisure increase with economic growth). Allowing for this growth can alter the assessment considerably. With growth in recreational benefits at 3% per annum, the required discount rate rises to 4.7% and with growth of benefits at 5% the project can be justified at a discount rate of 6.1%.[4]

These examples show how important discount rates can be in determining the viability of different kinds of investment in natural capital. High rates militate against investments that produce benefits after a long gestation period or where the benefits are spread out over a very long period. Often benefits are ignored when they do not generate monetary flows, which can also increase the rate of discount needed to justify the investment. In this case, however, the problem can be addressed by including the non-monetized benefits. Other issues not so easily addressed are how to deal with uncertainty, how to account for future growth in benefits and how to allocate a limited amount of capital for different projects when the proposed discount rate is too low. These questions are addressed in the following sections. The next section looks at the theoretical basis for choosing the discount rate the latest position on what is the appropriate rate for investments in natural capital. The following section discusses current guidelines and actual practice in discounting when valuing natural capital. The section titled "Applications to valuing different forms of capital" examines discounting in the context of specific ecosystems and the section titled "Conclusions and recommendations" offers some conclusions.

Determining the rate of discount

Alternative views of how discount rates should be determined

Arrow et al. (1996) distinguish between two fundamentally different approaches to the choice of a discount rate to compare consumption in different time periods: the *prescriptive* approach and the *descriptive* approach. The prescriptive approach to discounting begins by asking how trade-offs between present and future generations *should* be made. The descriptive approach, by contrast, begins by asking what choices involving trade-offs across time do people *actually* make? (Arrow et al., 1996: 129).

The distinction between these two approaches can be better understood by firstly considering the general expression of the social discount rate:

$$d = \rho + \theta g \tag{1}$$

Where d is the discount rate, ρ is the rate of pure time preference, θ is a measure of the rate in which consumption increases welfare (referred to technically as the elasticity of the marginal utility of consumption), and g is the growth rate of per capita consumption.[5]

The equation says that the discount rate is made up of two terms:

1   The pure rate of time preference (PRTP), (ρ) as a measure of individual or societal 'impatience'. If $\rho > 0$, current consumers care less about future utility (or welfare) than about today's welfare.
2   The growth in per capita consumption over time, expressed in units of utility (θg).

Among economists and philosophers there is little controversy about the second term of equation (1). If future generations are expected to become richer (a common assumption of most socio-economic scenarios underlying impact assessments of investment in natural capital), a future consumption bundle is worth less than an equivalent consumption bundle today. How much less depends on the growth of per capita consumption, g, and the elasticity of utility, θ, i.e., the marginal contribution of one extra unit of consumption to utility. Disagreements arise over what value of g can be assumed over long periods of time. Those favouring a lower rate of discount argue that society cannot sustain positive rates of growth of per capita consumption and so over the long term g should take a value close to zero.

Greater controversy focuses on ρ, the pure rate of time preference (PRTP). Advocates of the prescriptive approach to discounting argue that the PRTP should be zero or close to zero, because a higher rate could leave future generations short-changed, without the possibility of redress through intergenerational transfers. Advocates of the descriptive approach, however, argue that the PRTP should be inferred from actual (savings) decisions people make (revealed preference), and that a PRTP of zero, for example, is incompatible with current rates of savings in developed and developing countries (Arrow et al., 1996).

The debate regarding the value of ρ in selecting the social rate of discount is not new. Frank Ramsey (Ramsey, 1928), arguably the 'father' of the debate on discounting, stated that taking a value greater than zero was ethically indefensible. Of course, individuals may apply a higher pure rate of time preferences in making their decisions, as they face greater risks as individuals, including their possible death over the decision period. Society, however, can be seen as 'infinitely lived', or close to it. More recently, Stern has argued that the only reason for taking a positive value for the social pure rate of time preference is the "exogenous possibility of [human] extinction". On this basis he proposes a value of 0.01.

The prescriptive approach to discounting advocates a lower rate of discount than the descriptive approach. The growth rate of per capita consumption, g, depends on the underlying socio-economic scenario, and may differ between countries and time periods. The values of g in the Stern report and in the most widely used climate change models range between 1.5% and 2.0%, based on the experience of the 19th and 20th centuries. With a high g, discounting the future is justified by the assumption that those living in the future will be better off than those living today. World median household income today is about $10,000 (Phelps and Crabtree,

2013). At a 2% growth rate, real household median income in the year 2100 will be about $54,000, or 5½ times higher than today. At the 1.3% income growth rate used by Stern, it would be $35,000, or 3½ times higher. Even if climate change were to cut future incomes by 20% (Stern, 2007) people would still be on the order of 3 times to 4and a half times better off. This is the main argument for not taking aggressive action on global warming. Why should those living today sacrifice for the fabulously rich people living in the future?

One argument against taking such a rate of growth for a long time into the future is that the past experience of growth may not continue. Some take the view there are good reasons to believe it will slow down (Piketty, 2014). The other argument is that GDP growth does not translate into well-being through the assumed utility function used in such analysis. Such a view applies of course to the use of monetary valuation in all areas of public decision-making.[6]

Estimates of the elasticity of marginal utility, θ, usually range between 1 and 2. This implies that the marginal utility of consumption drops by 1% or 2 % when the level of consumption increases by 1%. The higher the value of θ, the greater is the weight given to a poor person's income relative to a rich person. Thus, if future generations are going to be richer than the present generation on account of economic growth, their utility is discounted more highly. Of course, if future generations were to be poorer, the same value of θ would imply their utility holds a greater weight. Some economists have attempted to infer a value of θ from government decisions relations to intra- and inter-generational transfers and have presented values within the range of 1–2 (Stern, 1977).

As an application of these principles, consider the rate of discount to be applied to evaluating programmes for climate change mitigation. In his assessment of climate change damage, Cline (1992) uses a PRTP of zero, a rate of growth of global per capita consumption (in the very long term) of 1%, and an elasticity of marginal utility of 1.5. Using Equation (1), this results in a discount rate of 1.5 % (Cline, 1992: 255). Nordhaus (1994), on the other hand, postulates a PTPR of 3%, a growth rate of per capita consumption of 3%, and an elasticity of marginal utility of 1, resulting in a discount rate of 6% (Nordhaus, 1994: 124). More recently Stern (2007) has assumed a PRTP of around 0.1% and a long term growth rate of zero resulting in a discount rate of 0.1%. The result is that with Stern's assumptions a much higher value of one ton of CO_2 released today is generated, because the future damages are hardly discounted at all. His damage estimate comes out at €85 (US$119) per ton of CO_2, compared to estimates that are an order of magnitude lower if we take discount rates of around 3%.

Arguments for changing the discount rate on grounds of uncertainty

The calculations of the discount rate presented previously assume that the latter does not change over time. But there is no reason for this to be the case and in particular there are a number of scholars who have presented reasons why the discount rate should decline with time. What follows is a short summary of these arguments.

Hyperbolic discounting (HD)

There is evidence to suggest that individuals and societies do not discount the future at a constant rate but rather that they adopt a 'hyperbolic' path (Kim and Zauberman, 2009; Settle and Shogren, 2004). To illustrate this, consider the following: an individual is faced with two choices: (1) postponing consumption for one year from now and (2) deferring an equal amount of consumption for one year from year 50 to year 51. Most individuals are likely to respond differently to these two choices. While postponing the consumption right now might mean a lot, postponing it for an equal amount of time in 50 years from now might not. In other words, the weight placed on an extra year in the future is declining with time. However, the standard formula for constant discounting gives the same value to both types of postponement.

Hyperbolic discounting means that the discount *factor* declines as a hyperbolic function of time. A general function proposed by Loewenstein and Prelec (1992) is as follows:

$$d_t = \frac{1}{(1+kt)^{h/k}} \tag{2}$$

Estimated values of the parameters are around 4 for k and around 1 for h. Figure 11.1 shows the values for hyperbolic discounted values of one unit against constant values discounted at 5%. The decline is very sharp initially and then tapers off, so that the values after a number of years are higher with this form of discounting than with constant discounting.

Revealed preference theory takes existing human preferences as the starting point for economic analysis. The evidence from behavioral economics shows quite clearly that actual choices about the future exhibit hyperbolic discounting (Gowdy et al., 2013; Laibson, 1997). People discount the immediate future at a higher rate

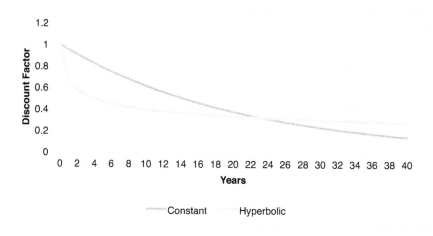

Figure 11.1 Hyperbolic vs constant discount rates.
Source: Author's computation

than the medium or distant future so that present discounted values flatten out after a period of time.

The main problems with hyperbolic discounting are:

1   Time inconsistency (decisions taken today for some time in the future will not look correct when that future time becomes the present – see what follows)
2   Very high discount rates in the immediate years, which are not congruent with actual rates expressed by individuals or societies

Discounting and uncertainty

Weitzman (1998, 2001) has made the point that if the parameters of equation (1) are not known with certainty, society should take a weighted discount rate based on its ignorance. This weighted discount factor is also called the Certainty Equivalent Discount Factor (CEDF). The result of taking an average of different discount rates is that over time the lowest rate prevails. This can be seen in Table 11.1, where the discount rate is assumed to be 1%, 3% or 5%, but we do not know which, and each has an equal probability of being right. The Certainty Equivalent Discount Factor is simply the sum of the individual discount factors, each multiplied by one-third. This simple process results in a discount factor with a certainty equivalent discount rate that is less than the simple average (which would be 3%) and that declines over towards 1%.

Furthermore, based on a survey of more 1,700 economists as to what the rates should be, Weitzman found that the distribution of the results looked like a gamma frequency distribution, rather than one with equal probabilities shown in Table 11.1. In that case the value of the discount rate over time is given by d_t.

$$d_t = \frac{\mu}{(1+t\sigma^2 / \mu)} \tag{3}$$

Where μ and σ are respectively the mean and standard deviation n of the gamma distribution, with values of around 4.1% and 3.1% respectively.

Table 11.1 Declining discount rate based on uncertainty

Discount Rates	Discount Factors in Period t				
	10	*50*	*100*	*200*	*500*
1% (p = 1/3)	0.905	0.608	0.370	0.137	0.007
3% (p = 1/3)	0.744	0.228	0.052	0.003	0.000
5% (p = 1/3)	0.614	0.087	0.008	0.000	0.000
Certainty Equiv. Discount Factor CEDF	0.754	0.308	0.143	0.046	0.002
Certainty Equiv. Discount Rate	2.86%	2.38%	1.96%	1.55%	1.22%

Source: Author's computation

Other researchers have also analysed the implications of not knowing the discount rate of the parameters of equation (1). Gollier (2002a, 2002b) examined lack of knowledge about θ and *g* in that equation. It turns out that if *g* is not going to decline in the medium to long term the discount rate declines only if θ declines with income. Recall that θ is the aversion to intergenerational inequality. It can also be interpreted as a measure of risk aversion. Taking the latter interpretation, however, one cannot argue that it declines with income and so the case for declining rates is not clear.

Newell and Pizer (2003) follow a different path from that of Weitzman. For them, future rates decline because of dynamic uncertainty about future events, not static disagreement over the correct rate, or an underlying belief or preference for deterministic declines in the discount rate. They take 200 years of US interest rates to quantify the uncertainties. Under a random walk model about movements in interest rates, the certainty equivalent rate falls continuously from 4% to 2% after 100 years, to 1% after 200 years, and to 0.5% after 300 years. At horizons of 400 years, the discounted value increases by a factor of over 40,000 relative to conventional discounting. Similar results are found when using a different model (the mean-reverting model) although to a lesser extent. The authors believe that the random walk model is more credible because the mean rate over the distant past is less informative than the recent past when we forecast at any horizon in the future. Thus, they get a similar result to Weitzman, albeit by a different route.

Declining discount rates by other means

Finally some researchers have argued for declining rates over time on grounds of equity and sustainability. Chichilnisky (1997) introduces two axioms for sustainable development, which, in combination, require that neither the present nor the future should play a dictatorial role in society's choices over time. The implications of her formula for discounting is that the future will be discounted in a conventional manner in the near future, but after a point – the so-called 'switching date' – remaining effects will not be discounted at all (i.e. at a zero rate).

Li and Lofgren (2000) consider two individuals with identical preferences in all areas except one: they have different values for ρ, the personal rate of time preference (see equation [1]). The overall societal objective is to maximize a weighted sum of utility for both of them. The result of this weighting practice is similar to that of Weitzman's discounting – the individual with lower discount rate is given dominant weight as time goes by and the collective discount rate is declining.

Conclusions on declining discount rates

In summary, this work has provided a compelling case for using declining discount rates, especially over the kind of long periods we are often looking at when deciding on policy related to investment in natural capital. The main objection to declining discount rates is time inconsistency. 'Time inconsistency' refers to the situation where plans made at one point in time are contradicted by later behaviour. To illustrate this the following example is taken from Guo (2004).

> If *A* prefers one dollar today to two dollars tomorrow, but prefers two dollars on the 51st day to one dollar on the 50th day and designs a consumption plan accordingly, but when the 50th day comes, *A* might decide to consume the one pound on that day instead of waiting until the 51st day.
>
> Likewise, if a government decides to use high discount rates for the near future but lower ones for the far future, the immediate large spending will be easily justified. However, when later governments review the policy, they may conclude that this earlier policy was not optimal and decide to increase the discount rate again, which will lead to higher consumption than planned.
>
> If used 'naively', therefore, declining discount rates, especially hyperbolic discounting (because of its problem of very high initial discount rates), may lead to a collapse of natural resources. This has been shown by Hepburn (2003). The main reason for time inconsistency is that governments cannot commit their future counterparts- reviewing policies is almost certain and legitimate.

Heal (1998) has also shown that almost all types of declining discount rates result in time inconsistency. The answer in the view of the present author is not to reject the use of declining discount rates but to maintain as much consistency as possible in the way in which they are used. Decisions about investments with very long term impacts should be carried out under lower rates than those with shorter term impacts. Of course, no one can ever commit future governments to a particular approach but if the basis of the present one has broad agreement, the chances of it being sustained are that much greater.

Current practice is moving in the direction of adopting declining discount rates. The biggest attempt has been made by the UK treasury in its Green Book (2003), where the recommended social discount rates are indeed declining with time. This is very much based on the analysis of uncertainty and follows closely the modelling of Newell and Pizer (2003). Rates proposed are given in Table 11.2. Note these refer to the real rate – i.e. net of inflation.

In a similar fashion, France decided in 2004 to replace its constant discount rate of 8% to a 4% discount rate for maturities below 30 years, and a discount rate that decreases to 2% for larger maturities.[7] Finally, the Office of Management and Budget of the US government both recognizes the possibility of declining rates (see appendix D of US, 2003). It should be noted, however, that for most public

Table 11.2 UK treasury recommended discount rates

Period of Years	0–30	31–75	76–125	126–200	201–300	300+
Discount Rate	3.5	3.0	2.5	2.0	1.5	1.0

Source: HM Treasury, UK, 2003

investment projects these lower rates will not kick in, as the time period over which they are appraised is usually 30 years or less.

Social rates of discount and individual rates of discount

The discussion so far has been on what the social rate of discount should be in valuing investments or in arriving at a value for an asset (e.g. some form of natural capital) that does not have a price. In the latter case, the value is determined as the *NPV* of the flow of services provided by the asset, with the present value calculated using the social rate of discount. Individuals and firms, however, do not apply such a rate when deciding on their investments or savings. In the case of individuals, they often borrow at very high rates of interest, implying a correspondingly high discount of the future. For firms, the key factor is the return on capital they can receive if they invested the same amount of money in some other asset or project, with the same degree of risk as with the asset being valued. Such rates (adjusted for inflation) can be very high, especially in developing countries.[8] Typical rates are in the region of 10% and above, depending on the risks involved.

How then, does one reconcile these two different rates? A common position taken by governments is to apply the social rate for investments and capital valuations in the public sector and leave the private sector to apply whatever rate it considers appropriate for its decisions. This is a workable solution in most circumstances, except that some private decisions involve investments and valuations of natural capital (such as minerals), which entail some use of natural capital that is not private. For example, consider a mine, which extracts a mineral. The value of the company owning the mine will reflect the expected present value of the flow of income, discounted at the market rate. However, the operations may be damaging a water body through disposal of tailings and one needs to value the change in the value of that body as well. The private valuation will only include a value based on any obligations imposed on it in terms of remediation or protection. The private value of the obligations at the private rate of discount will not be the same as the social value at the social rate – the latter being typically higher than the former. The only way to address this is through an additional valuation of the non-market impacts of private operations. The issues raised in relation to this are discussed further in the next section.

A related question is whether a lower rate of discount should be applied for investments in natural capital? For example, Hoel and Sterner (2007) show that if the relative prices of environmental goods increase over time (a likely prospect as they become scarcer) the effect of discounting would be offset, possibly even reversed. The relative price effect also lowers the discount rate itself, depending on the rate of decline in environmental quality, the elasticity of substitution between environmental and other goods, and the ability of technology to respond to increasing scarcity of environmental resources. The first effect of scarcity is relatively simple to account for: in the examples in the Introduction we saw the effects of growing environmental values on the rate at which projects could be justified. The second impact of scarcity work by reducing the discount rate through its impacts on the underlying growth rate (variable g in equation [1]). It is difficult, however, to see how making g endogenous can work at the practical level. A more workable approach would be to take bench-mark values of *g* that broadly reflect the same future for *all* projects.

Discounting practices in the current literature

In this section the approaches taken to discounting in international guidelines or doc-uments are discussed and compared to the theoretical review of the previous section. The documents considered are the SEEA central framework (European Commission et al., 2014), the TEEB studies and the World Bank work in capital accounting.

Discount rates as discussed in the SEEA (2012) guidelines

The System of Environmental-Economic Accounting 2012 (SEEA) Guidelines presented by the UN discusses the use of discount rates as a key component of the *NPV* approach. There is an entire annex devoted to it (Annex A5.2).

The guidelines start by noting the distinction between two types of discount rates as has been presented previously: individual discount rates and social dis-count rates. An individual discount rate entails consideration of preferences from the perspective of an individual consumer or firm; it is directly related to the prices for goods, services and assets presented to that individual. In addition, the prefer-ences are generally considered within the normal decision-making timeframes of an individual consumer or firm. Finally, the discount rate relevant to an individual consumer or firm needs to take into account the likelihood of earning interest (or, more generally, a return) so that consumption can be undertaken in the future.

A social discount rate reflects the time and risk preferences of a society as a whole. Unlike individuals, societies must consider future generations to a greater extent and must also balance the benefits accruing to different sections of society in current and future periods (i.e., the distribution of income and consumption). In addition, the risks associated with earning returns are far more dispersed and balanced at a societal rather than individual level and therefore the compensation for risk will usually be lower for a whole society. Often, social discount rates are applied in the context of a government in relation to its decision making on behalf of a society.

In valuing assets for the capital accounts, the objective is to calculate the *NPV* for assets that do not have a market value. In some cases, these assets are privately owned, in which case the SEEA argues that the private rate of discount for that assets, suitably adjusted for risk, "is the most appropriate discount rate in respect of aligning with the market price valuation principle used in the SEEA" (European Commission et al., 2014: 258).

It offers three ways of estimating this rate. One is to use the cost of funds to the firm or industry whose asset is being valued. This can be difficult when the costs vary across firms or when financial markets for the sector are not well developed, as may be the case in some developing countries.

The second way is to define the rate of return as equal to the total operating surplus of the firm or industry divided by the value of the stock of produced and apply that as the discount rate on natural resources. By its very construction, this rate will overstate the rate of return, since the denominator (the value of the stock of produced assets) excludes the value of natural resources. At the same time, this rate of return does take into account the returns accruing to the specific activity and hence the associated risks.

The third approach begins with the assumption that the rate of return on produced assets should be equal to an external rate of return that the firm would have received if it had invested in alternative assets. This rate is then assumed to apply also to natural resources. Since this rate of return takes into account investment in a broader range of assets across the economy, the industry-specific risks of investment are less likely to be accounted for.

SEEA argues that none of these relatively direct methods generates a discount rate that completely measures the desired concept but suggests a comparison of both rates may yield useful information. In particular, a useful approach may be to take a general, external rate of return as a base rate and to adjust it utilizing industry-specific information to account for specific investment risk. Adjustments may be made on the basis of relative financing costs or of the relative difference in the return to produced assets in the target industry compared with an economy-wide return to produced assets.

As far as the social discount rate is concerned, the discussions are similar to those presented in the previous section, though including a much shorter review of the issues arising with hyperbolic discounting or declining discount rates. At the end, however, the guidelines do not appear to recommend the use of a social rate. The final statement declares:

> It is recommended that a discount rate be determined that is consistent with the general approach to valuation in the SEEA and the SNA, i.e., consistent with valuation at market prices. This suggests the choice of an individual discount rate that reflects the return needed by those undertaking an activity to justify investment in that activity. Consequently, the relevant rate should be descriptive and, ideally, should include any activity-specific risks.
>
> (Page 231)

It further states that:

> Because judgements are required regarding societal preferences, it is not recommended that prescriptive approaches to the determination of discount rates be used for the purposes of official statistics.
>
> (Page 232)

Finally, a recommendation is made that some sensitivity analysis be carried out for the selected rate in complying with the valuation of environmental assets.

While one can appreciate the concern in using social discount rates, it is hard to justify the position taken in valuing environmental assets. Many of these assets generate returns over very long periods and as was shown in the first section, a high rate of discount places very little value to returns after about 50 years. When the valuation involves a *change* in the flow of services, such as carbon sequestration, the use of market-based rates will result in very low values that are not socially defensible. Thus, there is a case for using social discount rates to value such assets. It is possible to get political agreement on such rates, as has occurred in the United Kingdome, France and the United States.

This criticism of the SEEA is endorsed by some follow-on work undertaken by the European Commission, in collaboration with the OECD, UN and the World Bank (European Commission, 2013). It states:

> The SEEA Central Framework discusses discount rates and concludes that for the purpose of alignment of SEEA values with the SNA it is necessary to select marginal, private, market based discount rates in NPV calculations. This may not be considered appropriate for ecosystems as a whole whose value may be considered not properly reflected at the margin (European Commission et al., 2014: 155)

Discount rates used in the TEEB studies

TEEB does not present much theoretical discussion on the choice of the discount rate (UNEP, 2010). It notes that calculating natural capital involves discounting and reports it used a range of rates from 1% to 4% for this purpose. The justification was that such rates reflected an ethical position rather than a purely economic one. It reminds the reader that using a 4% rate means nature's benefits to our grandchildren are valued at one-seventh of what the current generation gets get from the planet. Even using a 1% discount rate is implying that a grandchild still only deserves two-thirds, and not more, of what the current generation currently get from nature.

Using these discount rates, TEEB concluded that the natural capital being lost was in the order of between $2.0 trillion and $4.5 trillion every year.

It is precisely as a response to the concerns expressed in the previous statement from the TEEB summary report that the research community has considered the concept of declining discount rates. In the short term, higher rates are justified on the grounds that they reflect societal preferences, but equally these same preferences point to lower rates for longer periods.

Discount rates in the World Bank wealth of nations and WAVES studies

The earlier World Bank study on Wealth of Nations (World Bank, 2006), which preceded the WAVES project, used the *NPV* method to value both physical and natural assets. Additionally, for assets that had a market value, the net price was also used. The net price method simply applies the price today net of extraction costs to the remaining stock to value that stock. So for example if the price of asset is $1 and its extraction cost is $0.1, and the stock consists of 1,000 units, its value is taken as $0.9 multiplied by 1,000, giving a total value of $900. Such a value but is considered as less accurate than the *NPV* method, which has superseded it.

For natural assets such as fisheries and forests, the rate used for most calculations was 4%, based on an estimate of the social discount rate made in earlier work by Pearce and Ulph (1999). The authors acknowledge that such a rate is likely be too low for fast-growing economies such as China, while being high for slow-growing economies in Sub-Saharan Africa. They decided upon a single discount rate for all countries in order to facilitate comparisons.

For physical assets, no discount rate was necessary when the asset had a market price, assumed to represent the discounted flow of future services from the assets. A separate calculation was made for the total capital stock based on the accumulated genuine savings from past output, discounted at a rate of 5%. The justification was that this represents the marginal product of capital and previous work had used a rate of 5% for that parameter.

The WAVES programme does not go further on the issue of discounting. A search of its publications reveals little on the subject, except in the Working Paper on Designing Pilots of Ecosystem Accounting (World Bank, 2014), where it essentially endorses the SEEA approach stating:

> The capacity of ecosystems to generate services needs to be valued based on the flow of present and discounted future benefits provided by the ecosystem, using an appropriate discount rate aligned with the SNA.

(Page 53)

Reconciliation between different practical approaches and the emerging theoretical literature

The literature on recommended and actual practice in discounting reveals some concerns. The theoretical literature suggests that for natural capital, it is desirable to use a social rate of discount, which could be represented by declining rates but which is in any case below the market rate. The UN-based SEEA guidelines recommend the use of a market-based rate, though that has been criticised in subsequent work. The World Bank in its most recent comprehensive natural capital assessment used a rate of 4% for most natural capital and 5% for minerals, acknowledging this to be a compromise between the rates appropriate in different countries but aiming to capture a social rate. Likewise, TEEB used a rate of 1–4% with the same objectives. Further work is needed to determine the appropriate rate for different ecosystems, and the following section explores this.

Applications to valuing different forms of capital

Introduction to Ecosystem Services (ESS) and their values

This section considers the biomes, maps the main ESS to each biome, and discusses the valuation issues that arise in valuing each of the ESS. It also discusses issues arising in the valuation of minerals. The ESS to be considered in the valuation of natural capital are given in Table 11.3.

The different ESS in each biome can be seen from the information collected by De Groot et al. (2012). They identified more than 1,600 studies over the period 1960 to 2008 and extracted 655 data points that could be used to calculate the flow of services in terms of international dollars per hectare per year.[9] Studies conducted

Table 11.3 List of different ecosystem services

Type of ecosystem service	
Provisioning services	**Regulating services**
Food and fibre	Air quality maintenance
Fuel	Climate regulation (e.g. temperature and precipitation, carbon storage)
Biochemicals, natural medicines, and pharmaceuticals	Water regulation (e.g. flood prevention, timing and magnitude of runoff, aquifer recharge)
Ornamental resources	Erosion control
Fresh water	Water purification and waste management
Cultural services	Regulation of human diseases
Cultural diversity, spiritual and religious values, educational, inspiration, aesthetic values, social relations, sense of place and identity	Biological control (e.g. loss of natural predator of pests)
Cultural heritage values	Pollination
Recreation and ecotourism	Storm protection (damage by hurricanes or large waves)
Supporting services	Fire resistance (change of vegetation cover lead increased fire susceptibility)
Primary production	Avalanche protection
Nutrient cycling	Other (loss of indicator species)
Soil formation	

Source: MEA (2005)

Table 11.4 Major biomes used in the ecosystem valuation literature

Biome (marine/aquatic?)	Biome (terrestrial?)
Marine (Open Oceans)	Freshwater (Rivers/Lakes)
Coral Reefs	Tropical Forests
Coastal Systems	Temperate Forests
Coastal Wetlands	Woodlands
Inland Wetlands	Grasslands

Source: De Groot et al. (2012)

in different currencies were converted into US dollars using purchasing power parity (PPP) exchange rates and account was taken of inflation between the year of study and the standardised year, which was 2007.

Table 11.5 provides the main results that emerge from this literature review. The results show significant values accruing to services in the different biomes, ranging from a maximum of US$352,249/ha/yr for coral reefs to a minimum of US$491/ha/ yr for marine areas. The most prominent services in descending order were regulating, followed by cultural, provisioning and habitat. Within the regulating services, ecosystems provide an important service of waste treatment and erosion prevention. The biomes in which they are found are listed in Table 11.4.

Table 11.5 Summary of monetary values for each service by biome (in Int. $/Ha./Yr. 2007 Price level)

		Marine	Coral Reefs	Coastal Systems	Coastal Wetlands	Inland Wetlands	Rivers and Lakes	Tropical Forests	Temperate Forests	Woodlands	Grasslands
Provisioning Services Total		102	55,724	2,396	2,998	1,659	1,914	1,828	671	253	1,305
1	Food	93	667	2,384	1,111	614	106	200	299	52	1,192
2	Water				1,217	408	1,808	27	191		60
3	Raw Materials	8	21,528	12	358	425		84	181	170	53
4	Genetic Resources		33,048		10			13			
5	Medicinal Resources					99		1,504			
6	Ornamental Resources		472		301	114				32	1
Regulating Services Total		65	171,478	25,847	171,515	17,364	187	2,529	491	51	159
7	Air Quality Regulation							12			
8	Climate Regulation	65	1,188	479	65	488		2,044	152	7	40
9	Disturbance Moderation		16,992		5,351	2,986		66			
10	Water Flow Regulation					5,606		342			
11	Waste Treatment		85		162,125	3,015	187	6	7		75
12	Erosion Prevention		153,214	25,368	3,929	2,607		15	5	13	44
13	Nutrient Recycling				45	1,713		3	93		
14	Pollination							30		31	
15	Biological Control					948		11	235		
Habitat Services Total		5	16,210	375	17,138	2,455		39	862	1,277	1,214
16	Nursery Service			194	10,648	1,287		16		1,273	
17	Genetic Diversity	5	16,210	180	6,490	1,168		23	862	3	1,214
Cultural Services Total		319	108,837	300	2,193	4,203	2,166	867	990	7	193
18	Esthetic Information		11,390			1,292				7	167
19	Recreation	319	96,302	256	21,930	2,211	2,166	867	989	7	26
20	Inspiration					700					
21	Spiritual Experience		1,145	21							
22	Cognitive Development			22					1		
	Total Economic Value	491	352,249	28,918	193,844	25,681	4,267	5,263	3,014	1,588	2,871

Source: De Groot et al. (2012)

Note: Coastal systems include estuaries, continent shelf areas and sea grasses but exclude wetlands like tidal marshes, mangroves and saltwater wetlands

Values of the different biomes and minerals

The different biomes are the basis on which the valuation of natural capital is performed. Each biome contains a number of ESS that need to be discounted to the present to obtain an *NPV*. The sum of these *NPV*s gives the total value of that biome in terms of natural capital. In the following sections, each of the biomes are considered in turn. Climate regulation ESS, however, are discussed separately as they are relevant for all biomes.

Marine biomes

In the marine system the key ESS apart from climate regulation (which is addressed later) are food provision in the form of fisheries and aquaculture, and recreation. Regarding fisheries, the main difficulties when calculating NPV include estimating future yields and accounting for changes in future prices. There is little guidance on how to estimate future yields. The dynamics of fish stocks are complex to model and require integration of biological and economic modeling and external factors such as climate change and overfishing, which may deplete stocks considerably in some locations. Given the changes in stocks, catch rates also depend on effort (number of boats, time spent fishing, etc.), which is an economic decision variable. This of course will have an effect on the yield and on the net income from the activity.

In a recent and comprehensive review of the marine fishery sector in the context of greening the economy, UNEP (2011) does not attempt to provide an estimate of the capital value of different marine waters. Likewise, earlier efforts at measuring the natural capital stock such as World Bank (2006) does not deal with this sector. The UNEP report states that currently, fisheries yield a negative net value – of US$26 billion a year, when the total cost of fishing (US$90 billion) and non-fuel subsidies (US$21 billion) are deducted from the total revenues of US$85 billion that fishing generates. It further notes that a transformation of practices would allow the world to rebuild overexploited and depleted fish stocks. This would involve greening the sector by reorienting public spending away from subsidies and towards strengthening fisheries management and a reduction of excess capacity through de-commissioning vessels and equitably relocating employment. An investment of US$100–300 billion would reduce excessive capacity, and result in an increase in fisheries catch from the current 80 M tons a year to 90 M tons in 2050, despite a drop in the next decade as stocks recover. The present value of benefits at a 5% discount rate from greening the fishing sector is about 3 to 5 times of the necessary additional costs.

Such an analysis is a valuable contribution to the debates around on greening the economy and values investments and changes in policies; however it does not value the stock of natural capital. The latter should be possible given the data collected, but further work would be required. The appropriate discount rate for the calculations is perhaps better set at a lower value than 5% and it may be worth considering the use of declining rates for the exercise. Most importantly, however, is to get the physical estimates of possible catch rates under current practices and estimates of future prices as correct as possible.

The other major ESS marine systems provide is recreation, where the major challenge is to accurately estimate future demand and willingness to pay for recreational uses. As discussed previously, the key to this is to work with scenarios that capture likely future growth in individual disposable income in different parts of the world, as well as changes in the population demanding such services.

Forests

Forest biomes are categorized in Table 11.6 into tropical and temperate forests and woodlands. Apart from carbon sequestration values accruing from these forms of natural capital, the other main sources of ESS are raw materials (timber) and a multitude of non-timber services, such as nutrient recycling, erosion prevention, recreation, food products and genetic materials. In the work done to date on valuing forests as a form of capital all these non-timber resources and services have been grouped as a single entity.

For the valuation of timber, the NPV of rents from roundwood production has been estimated. This requires data on roundwood production, unit rents, and the time to exhaustion of the forest (if unsustainably managed). All these components raise some difficult problems and calculating the rent entails complexities. Theoretically, the value of standing timber is equal to the discounted future stumpage price received by the forest owner after taking out the costs of bringing the timber to maturity. In practice, stumpage prices are usually not readily available, and unit rents have been calculated as the product between a composite weighted price times a rental rate. This rate needs to be estimated region by region and can be quite onerous. In the World Bank (2006) study these rates were estimated using available studies and consultation with World Bank forestry experts. Likewise, estimates were made for future prices based on detailed studies of the timber market.

Another issue to consider is the period of time over which the rent should be calculated. In this case, two estimates were made: first, the roundwood harvest was compared to the net annual increment to the stock. If the harvest is smaller than this increment, that is, the forest is sustainably harvested, the time horizon was set at 25 years. If the roundwood harvest was greater than the net annual increments, then the time to exhaustion was calculated. The time to exhaustion is based on estimates of forest volume divided by the difference between production and increment. The smaller of 25 years and the time to exhaustion is then used as the resource lifetime. The use of a 25 year horizon was justified on the grounds that (1) at the 4% discount rate, values after 25 years play a relatively small part in the total value and (2) estimates of prices beyond that period have very high uncertainties.

For the non-timber services, those considered in the World Bank study were minor forest products, hunting, recreation, watershed protection, and option and existence values. A review of non-timber forest benefits in developed and developing countries revealed that returns per hectare per year from such benefits vary from US$190 per hectare in developed countries to US$145 per hectare in developing countries (based on Lampietti and Dixon 1995 and on Croitoru and others

2005, and adjusted to 2000 prices). The study assumes that only one-tenth of the forest area in each country is accessible, so this per hectare value is multiplied by one-tenth of the forest area in each country to arrive at annual benefits. Non-timber forest resources are then valued as the net present value of benefits over a time horizon of 25 years.

The treatment of discounting here has so far been somewhat limited. A rate of 4% has been used and a time horizon of 25 years has been imposed. Regarding forests, it is at least worth looking at the implications of lower discount rates, possibly declining ones, and considering much longer horizons. These factors will be important when appraising a shift to a more sustainable management system as opposed to one that results in increased deforestation. It is also important to note that the values of non-timber benefits are incomplete and do not fully capture value aspects such as genetic diversity and existence value. Further work on these aspects, as well as the ways in which they vary over time is required.

Other biomes

Biomes classified in Table 11.5 as grasslands cover areas that are in part used for crops and as pasture and in some cases, as protected areas. For cropland and pastureland, the main service provided by the biome is the food products derived from the land. These are valued at the market prices net of costs of production. The discounted value of the stream of these net incomes is the value of the land from which the services are obtained. This raises similar problems as the valuation of forests: how data can be obtained on net incomes now and in the future. The World Bank (2006) valuation study used external estimates of the 'rental rate' – i.e. percentage of the price that is net income and combined it with data on prices from FAO. For future years prices were assumed constant and a horizon of 25 years was imposed, for similar reasons as those given in the case of forests. The discount rate used was 4%. As with the case of forests, these simplifications may be justified on practical grounds but they mask the problems of changing future prices; and the possibility of declining yields due to degradation in some locations or of increasing yields due to technological improvements in others.

In the case of protected areas, the best approach to estimate NPV would be to estimate the willingness to pay for visits to the areas, plus any existence values now and in the future. In practice, this is difficult to conduct for all sites that need to be valued, although given the recent work on ESS valuation, such an exercise should be possible. Instead, the World Bank study used the concept of opportunity cost to value these areas. The opportunity cost is the value of what could have been produced on the land if it had not been declared as protected. Typically, such areas would be used for pasture or growing crops and obtaining fuelwood. Some of these uses may be compatible with protected area status, in which case they can be used to measure those values. But mostly, cultivation or grazing is not permitted, in which case the valuation is a measure of the lost output.

Summary of issues arising in the valuation of biomes

This discussion demonstrates some of the complexities arising in the valuation of biomes as a form of natural capital. It is difficult, to provide estimates of the current values of ESS, though recent work has demonstrated that this is possible for many services. The next step is to value the future flow of services, which raises questions about the sustainability of the use of the biome at its present level, the impacts of external factors such as climate change on the level of services and changes in the demand for services, which may alter the prices associated with them. Furthermore, the choice of the discount rate is also an important factor and this has been treated quite cursorily in the literature. A single rate has been used, which is chosen as a rough approximation to the social rate and which can vary according to the kind of capital being valued. Furthermore, the valuation is applied for a limited time period of 20–25 years. Long term valuations do not seem to have been carried out and the issues of declining discount rates are not investigated. Some attention has been brought to issues of sustainability but only for forests and not for other biomes. Indeed, not all biomes have been valued for their natural capital, with marine ecosystems being a prominent example.

Minerals

Although not included in the list of ESS in Table 11.6, an important source of natural capital is the stock of minerals, which are non-renewable and therefore eventually exhausted. The World Bank (2006) identifies two difficulties in calculating the value of such natural capital. First there are no private markets for subsoil resource deposits to convey information on the value of these stocks. Second, the stock size is defined in economic terms – reserves are "that part of the reserve base which could be economically extracted or produced at the time of determination" – and, therefore, it is dependent on the prevalent economic conditions, namely technology and prices.

The method used for the valuation of such assets is the NPV described in this report. The value of a given mineral in a given region is estimated by the following formula:

$$V_t = \sum_{i+t}^{t+T-1} \pi_i q_i / (1+r)^{(i-t)} \tag{4}$$

Where $\pi_i q_i$ is the economic profit or total rent at time i (Πi denoting unit rent and qi denoting production), r is the social discount rate, and T is the lifetime of the resource.

The main problem with the application of the method is estimating future rents. Usually, the valuation is based on some simplification of how future rents are related to present rents (which can be estimated). One assumption may be that the rents are constant, another is that future extraction and future rents will be determined on an optimal extraction path, defined by maximising the discounted net present value of rents based on private discount rates. The World Bank used the assumption that rents will grow at a rate g, which is based on optimal extraction as defined previously, and which depends on the costs of extraction in the following way:

$$g = \frac{r}{1+(\varepsilon-1)(1=r)^T} \qquad (5)$$

Where ε is the elasticity of the cost of extraction with respect to quantity extracted and is estimated at being 1.15. The resulting estimate of V in (4) is then given by:

$$V_t = \pi_t q_t \left(1+\frac{1}{r^*}\right)\left(1-\frac{1}{(1+r^*)^T}\right) \qquad (6)$$

Where r^* can be seen as the effective discount rate and defined as:

$$r^* \equiv \frac{(r-g)}{(1+g)} \qquad (7)$$

That leaves an estimate of T to be made, which the World Bank did based on external estimates of the ratio of economically estimated reserves to current production. Data on these are available for the major minerals and countries. In the end, however, the World Bank limits the value of T to 20 to 30 on the grounds that estimates of rents beyond that period are more uncertain and that using a discount rate of 5% means that rents in the long term have a small impact on the total value.

As noted, the estimates of mineral capital are based on discount rates (i.e. the value of r) are about 5%. This parameter, however, needs further investigation. It is important to examine the impacts of lower and/or declining discount rates on the valuations obtained. Other issues that need to be investigated further are: (1) impacts of changes in technology that reduce the costs of extraction, and (2) the impact of increasing mineral rents on reserves and thereby on the value of T.

Carbon sequestration

The main questions arising in the valuation of carbon sequestration services are the values attached to carbon not released now and in the future. The value of carbon in this context is referred to as the Social Cost of Carbon (SCC).

These costs have been reviewed in some depth in the literature (DEFRA, 2007; Anthoff and Tol, 2013; US Government 2013). The values are based on the discounted damages arising from one tonne of CO_2 over the long term and therefore are sensitive to the discount rate adopted. The higher the discount rate, the lower will be the value attached to future damages and hence the lower will be the discounted present value of the damages. This discounted value also increases over time as damages rise with increased temperatures. The US Government Review of 2013 is arguably the most comprehensive recent assessment. Box 11.1 describes the elements in the calculation of the social costs of carbon from the different models that have been used and that are covered in this review.

Box 11.1   Elements in the Social Costs of Carbon (SCC)

As explained in the text SCC is calculated by running an Integrated Assessment Model (IAM) in which future economic output is estimated under different scenarios for emissions of GHGs. By running the model with given emissions scenario the discounted present value of output is obtained. Running the model again with a small increase in emissions in the current period a second discounted present value is obtained. Subtracting the discounted value in the second run from the value in the first gives an estimate of the damage caused by that small increase is arrived at. Dividing the damage by the change in emissions gives the SCC today. The same calculation can be made starting the model in 2020, 2030 etc., to get the SCC for that year.

The impacts of climate change taken into account vary from one model to another. Three major models are DICE, FUND and PAGE. All include the damage caused by sea level rise (SLR), agriculture and energy (higher demand for energy for cooling but less for heating). These also include additional costs of health treatment resulting from higher temperatures and extreme events. Models vary in the damage function they use (i.e. the link between emissions and climate change and between climate change and damages) and there is an element of arbitrariness about the functions. Elements not included in the models are:

1   **Incomplete treatment of non-catastrophic damages:** current IAMs do not assign value to all important physical, ecological and economic impacts of climate change, and it is recognised that even in future applications a number of potentially significant damage categories will remain non-monetised i.e. ocean acidification (not quantified by any of the three models), species and wildlife loss.

2   **Incomplete treatment of potential catastrophic damages:** damage functions may not capture the economic effects of all possible adverse consequences of climate change, i.e. **(1)** potentially is continuous 'tipping point' behaviour in Earth systems; **(2)** inter-sectoral and inter-regional interactions, including global security impacts of high-end warming, and **(3)** imperfect substitutability between damage to natural systems and increased consumption.

3   **Uncertainty in extrapolation of damages to high temperatures:** estimated damages are far more uncertain under more extreme climate scenarios.

4   **Incomplete treatment of adaptation and technological change:** models do not adequately account for potential adaptation or technological change that may alter the emissions and resulting damages.

Source: US Government (2013)

Table 11.6 SCC values in US$2007 per metric tons of CO_2

Discount Rate Year	5% Average	3% Average	2.5% Average	3% 95th Percentile
2010	11	33	52	90
2015	12	38	58	109
2020	12	43	65	129
2025	14	48	70	144
2030	16	52	76	159
2035	19	57	81	176
2040	21	62	87	192
2045	24	66	92	206
2050	27	71	98	221

Source: US Government (2013)

The sensitivity of SCC to the discount rates can be seen from Table 11.6, taken from the US government report. Three points should be noted in this table. First halving the discount rate increases the SCC by more than a factor of 4 in 2010 and by just under a factor of 4 in 2050. Second, for any given discount rate, the figures given in the columns 2–4 are averages. Because of the uncertainty in damages there is a distribution around the mean and the last column gives the 95th percentile values for a discount rate of 3%. These are about 3 times greater than the mean. Lastly, there is a substantial increase in SCC over time: between 2050 and 2010 the value per tonne of carbon rises by a factor of about 2.5 for a high discount rate and a factor of about 2 for a lower discount rate.

The calculations of the value of natural capital the carbon component needs to take these factors into account. The choice of the discount rate here is a matter of social policy but, given the arguments presented earlier, a rate close to 3% is more justifiable than one close to 5%.

Another question to consider is whether a discount rate should be applied to carbon. In a mitigation project, should reductions of GHGs in the future be valued less than reductions today? One could argue that this should indeed be the case, as the impacts of future reductions will be less. This is especially true of "sink" projects, some of which will yield carbon benefits well into the future.

Many estimates of the benefits of reductions in GHGs do not apply a discount rate to the carbon changes. Instead, they simply take the average amount of carbon stored or reduced over the project lifetime (referred to as flow summation) or take the amount of carbon stored or reduced per year (flow summation divided by the number of years). Where some discounting is applied it has tended to be at a very low rate (less than 1%, Boscolo et al., 1998). This, however, is not a robust approach; as Table 11.6 demonstrates, capturing carbon at different dates provides different values and one cannot simply add up future and present amounts to get a single value to which a single unit value is attached. If, however, this is the chosen method, then physical discounting of future carbon represents a combination of a social discount rate plus the effect of changes in the unit value of future versus present amounts of carbon sequestered and it is not clear what the right discount rate will be for the resulting combination.

Discounting carbon entails a number of complexities as can be seen from the following example (Sedjo et al., 1997). Suppose there is a forestry project that will sequester 2 tons/C/ha/year for 50 years. At the end of that period it appears that the forest will be cleared completely. In physical terms then, the 'project' has stored no carbon. Yet if one discounted the savings for years 1–49 and the release in year 50 at 4%, one would estimate a carbon 'value' of 29.4 tons. It could be argued that there is a small benefit from such an event because without it there would have been a small release every year for 49 years and this has been avoided. The issue then boils down to what value can be attached to the carbon released or captured at different points in time. If such valuation is being done and discounted appropriately, using the SCC given in Table 11.6, there is no need to also discount the emissions of carbon. But if the accounting in purely physical terms, as it is in some programmes, there is a case for discounting reductions according to when they are made. The appropriate rate would be the rate applied to climate damages, as summarized in the previous section.

Conclusions and recommendations

Regarding the valuation of any form of capital, the choice of the discount rate is a key factor, and natural capital is no exception. The higher the discount rate chosen, the lower the NPV of a given stream of benefits and the lower the value given to natural capital now. Why does this matter? One reason is that decisions involving the destruction of natural capital become easier to justify. Clearing a forest to expand agriculture or for growing biofuels is more likely to be favoured in a cost-benefit analysis. Another reason is that higher rates of exploitation of a non-produced mineral would be justified, as would unsustainable rates of exploitation of a renewable resource. If the future values are highly discounted, such practices will be judged as acceptable as the future losses are negligible and present gains high.

The discount rate for valuing natural capital can be either the private rate or the social rate. Private rates are almost always much higher than social rates and can be warranted on the grounds that they reflect the preferences of the people who own the resources. However, most natural capital is a public good, or includes many services that are considered public goods. Thus, social rates are more appropriate for valuing such forms of capital and we can look to the theoretical discussions on this issue for providing justification for use of social rates. Some of the key messages arising from these discussions indicate that perhaps social preferences are best represented by declining rates, so that services received in the short run are discounted at a higher rate than services received in the long run. Such a structure better captures social preferences on costs and benefits at different points in time than would a constant discount rate. It also better captures the impact of uncertainty in comparing flows at different points in time. Some governments (United Kingdom, France, the United States in part) have adopted declining discount rates to value costs and benefits of long term projects (over 30 years). The guidance literature on capital accounting, however, has not taken these onboard and still argues for the use of a private rate of discount, though this is position is increasingly being criticised.

The practical literature on valuing natural capital has dealt quite cursorily with the choice of the discount rate. A single rate has been used, which may vary from one form of capital to another and is chosen as a rough approximation to the social rate. Furthermore, it has been applied for a limited horizon of 20–25 years. This limitation is justified on the grounds that at typical discount rates, the benefits after 25 years play a small part in the total value and that estimates of other components of future value beyond that period have very high uncertainties. Long term valuations seem absent from practice and the issues surrounding declining discount rates have not been adequately investigated.

Carbon sequestration services are one aspect of discounting that has been discussed in some detail. The value of such services, associated with a number of types of natural capital, will depend critically on the choice of the discount rate. This relation is captured through the social cost of carbon (SCC) and estimates of this parameter have been used valuing the flow of sequestration services in various contexts.

In general, the work on natural capital has been wrestling with several other problems of valuation, especially obtaining sound data on the flow of services and their likely future values and has devoted less attention to the impacts of discount rates. This aspect, does, however, deserve more attention, and is well overdue.

Notes

1 Not all choices between the present and the future involve money. Sometimes it may be a landscape or some indicator of the quality of the environment. In using discount rates the assumption is made that choices between the present and the future can be converted into money values. It is also important to note that the discount rate applied here is a real rate. That is to say future payments are assumed to be in real terms, net of inflation. If there is inflation of, say 3% and the actual rate of discount is, say, 7%, then the real rate net of inflation is 4%.

2 The benefits over time are measured net of any costs incurred with the investment.

3 It is hard to envisage the situation after that and in practice the calculations are not much affected by curtailing the estimation with a horizon of 2100.

4 The growth in recreational benefits can also been seen as a way of reducing the effective discount rate for the project. If benefits grow at 3% for example, this is akin to reducing the actual discount rate by such a percentage. This point was noted a long time ago by Krutilla and the discussion can be found in Krutilla and Fisher (1975). In practice the impact is not quite equal to the growth in benefits on account of the different time periods over which benefits and costs apply.

5 θ can also be interpreted as society's aversion to inequality. The higher the value the greater the penalty attached to future generations if they are richer than present ones.

6 Some economists (Dasgupta and Mäler, 2000) have argued that the 'correct' value of g in (1) is the growth of total factor productivity (TFP), which is the growth of output less the weighted growth in manufactured, human and natural capital. In cases where

natural capital is being depleted TFP could be negative when account is taken of such losses. This could lead to negative discount rates. In practice, however, such rates have not been used either in project appraisal or the modelling work involving growth with natural capital, such as climate policy.

7 This should be interpreted as using a discount factor equaling $(1.04)^{-t}$ if the time horizon t is less than 50 years, and a discount rate equaling $(1.04)^{-30}(1.02)^{-(t-30)}$ if t is larger than 30.

8 The discount rates discussed in this report are all net of inflation and are referred to as the real rates as opposed to the nominal rates. If the nominal rate is 10% and inflation is 6% then the real rate of discount is 4%.

9 International or Geary-Khamis dollar is a unit of currency constructed to standardize money values by correcting money values across countries to the same purchasing power that the US dollar has at a point in time. This involved using PPP exchange rates as has been done in the table cited.

References

Anthoff, D., and R.S.J. Tol (2013). "The Uncertainty about the Social Cost of Carbon: A Decomposition Analysis Using FUND." *Climatic Change* 117: 515–530.

Arrow, K.J., W.R. Cline, K.-G. Maler, M. Munasinghe, R. Squitieri, and J.E. Stiglitz (1996). "Intertemporal Equity, Discounting, and Economic Efficiency." In *Climate Change 1995: Economic and Social Dimensions of Climate Change: Contribution of Working Group III to the Second Assessment Report of the Intergovernmental Panel on Climate Change*, ed. J.P. Bruce, H. Lee, and E.F. Haites, 125–144, Cambridge, MA: Cambridge University Press.

Boscolo, M., J.R. Vincent, and T. Panayatou (1998). "Discounting Costs and Benefits in Carbon Sequestration Projects'." Environment Discussion Papers, No 41, Harvard Institute for International Development, Harvard, Cambridge, MA.

Chichilnisky, G. (1997). "What Is Sustainable Development." *Land Economics* 73: 467–491.

Cline, W.R. (1992). *The Economics of Global Warming*, Washington, DC: Institute for International Economics.

Croitoru, L., P. Gatto, M. Merlo, and P. Paiero., eds. (2005). *Valuing Mediterranean Forests: Towards the Total Economic Value*, Rome: CABI Publishing.

Dasgupta, P., and K.-G. Mäler (2000). "Net National Product, Wealth and Social Well-Being." *Environment and Development Economics* 5: 69–93.

DEFRA (2007). *The Social Cost of Carbon and the Shadow Price of Carbon: What They Are, and How to Use Them in Economic Appraisal in the UK*, London: Economics Group, Defra. www.decc.gov.uk/assets/decc/what%20we%20do/a%20low%20carbon%20uk/carbon%20valuation/shadow_price/background.pdf.

De Groot, R., L. Brander, S. van der Ploeg, R. Costanza, F. Bernard, L. Braat, M. Christie, N. Crossman, A. Ghermandi, L. Hein, S. Hussain, P. Kumar, A. McVittie, R. Portela, L.C. Rodriguez, P. ten Brink, and P. van Beukering (2012). "Global Estimates of the Value of Ecosystems and Their Services." *Ecosystem Services* 1: 50–61.

European Commission, Food and Agriculture Organization of the United Nations, International Monetary Fund, Organisation for Economic Co-operation and Development, United Nations and World Bank (2014). System of Environmental-Economic Accounting 2012: Central Framework. Studies in Methods, Series F, No. 109. Sales No. E.12. XVII. 12.

Gollier, C. (2002a). "Time Horizon and the Discount Rate." *Journal of Economic Theory* 107: 463–473.

Gollier, C. (2002b). "Discounting an Uncertain Future." *Journal of Public Economics* 85(2): 149–166.

Gowdy, J. (2013). "Valuing Nature for Climate Change Policy: From Discounting the Future to Truly Social Deliberation." In *Handbook on Energy and Climate Change*, ed. R. Fouquet, 547–560, Edward Elgar Cheltenham, UK.

Guo, J. (2004). *Discounting and the Social Cost of Carbon*. Thesis for the Masters' Degree in Environmental Change and Management, University of Oxford, Oxford.

Heal, G. (1998). *Valuing the Future: Economic Theory and Sustainability*, New York: Columbia University Press.

Hepburn, C. (2003). "Hyperbolic Discounting and Resource Collapse." Oxford University Department of Economics Working Paper.

Hoel, M., and T. Sterner (2007). "Discounting and Relative Prices." *Climatic Change* 84: 265–280.

Kim, B., and G. Zauberman (2009). "Perception of Anticipatory Time in Temporal Discounting." *Journal of Neuroscience, Psychology, and Economics* 2: 91–101.

Krutilla, J., and A. Fisher (1975). *The Economics of Natural Environments: Studies in the Valuation of Commodity and Amenity Resources*, Baltimore: John Hopkins University Press.

Laibson, D. (1997). "Golden Eggs and Hyperbolic Discounting." *Quarterly Journal of Economics* 112: 443–477.

Lampietti, J., and J. Dixon (1995). "To See the Forest for the Trees: A Guide to Non-Timber Forest Benefits." Environment Department Paper 13. World Bank, Washington, DC.

Li, C.Z., and K.G. Lofgren (2000). "Renewable Resources and Economic Sustainability: A Dynamic Analysis with Heterogeneous Time Preferences." *Journal of Environmental Economics and Management* 40: 236–250.

Loewenstein, G., and D. Prelec (1992). "Anomalies in Intertemporal Choice: Evidence and an Interpretation." *Quarterly Journal of Economics* 107: 573–597.

MEA (2005). *Millennium Ecosystem Assessment: Ecosystems and Human Wellbeing: Current State and Trends*, Washington, DC: Island Press.

Newell, R.G., and W.A. Pizer (2003). "Discounting the Distant Future: How Much Do Uncertain Rates Increase Valuations?" *Journal of Environmental Economics and Management* 46: 52–71.

Nordhaus, W.D. (1994). *Managing the Global Commons: The Economics of Climate Change*, Cambridge: The MIT Press.

Office of Management and Budget, Circular A-94: Guidelines and Discount Rates for Benefit-Cost Analysis of Federal Program, Transmittal Memo No. 64. [www.whitehouse. gov/omb/circulars/a094/a094.pdf] on 1/31/03.

Pearce, D.W., and D. Ulph (1999). "A Social Discount Rate for the United Kingdom." In *Environmental Economics: Essays in Ecological Economics and Sustainable Development*, ed. D.W. Pearce, 268–285, Cheltenham: Edward Elgar Publishing.

Phelps, G. and S. Crabtree (2013). Worldwide, Median Household Income About $10,000: Country-level income closely related to Payroll to Population results. (www.gallup.com/ poll/166211/worldwide-median-household-income-000.aspx)

Piketty, T. (2014). *Capital in the Twenty-First Century*, Cambridge, MA: Harvard University Press.Ramsey, F. (1928). "A Mathematical Theory of Saving." *Economic Journal* 38(152): 543–549.

Sedjo, R.A., R. Neil Sampson, and J. Wisneiwski, eds. (1997). Economics of Carbon Sequestration in Forestry in Critical Reviews in Environmental Science, 27 (Special issue).

Settle, C., and J. Shogren (2004). "Hyperbolic Discounting and Time Inconsistency in a Native Exotic Species Conflict." *Resource and Energy Economics* 26: 255–274.

Stern, N. (1977). "The Marginal Valuation of Income." In *Studies in Modern Economic Analysis*, ed. M.J. Artis and A.R. Nobay, Oxford: Blackwell.

Stern, N. (2007). *The Economics of Climate Change: The Stern Review*. Cambridge: Cambridge University Press.

Treasury, H.M. (2003). *The Greenbook Appraisal and Evaluation in Central Government*, London: TSO.

UNEP (2010). *Our Planet: Natural Capital: The Economics of Ecosystems and Biodiversity*, Nairobi: UNEP.

UNEP (2011). *Fisheries: Investing in Natural Capital*, Nairobi: UNEP.US Government (2013). "Technical Support Document: Technical Update of the Social Cost of Carbon for Regulatory Impact Analysis Under Executive Order 12866." Interagency Working Group on Social Cost of Carbon, United States Government.

Weitzman, M.L. (1998). "Why the Far-Distant Future Should Be Discounted at Its Lowest Possible Rate." *Journal of Environmental Economics and Management* 36: 201–208.

Weitzman, M.L. (2001). "Gamma Discounting." *American Economic Review* 91(1): 261–271.

World Bank (2006). *Where Is the Wealth of Nations?*, World Bank Publications, Washington, DC.

World Bank (2014). *Working Paper on Designing Pilots of Ecosystem Accounting*, Washington, DC: WAVES Programme.

Index